Climate, Capitalism and Communities

Also available:

*Boomtown:*
*Runaway Globalisation on the Queensland Coast*
Thomas Hylland Eriksen

*Overheating:*
*An Anthropology of Accelerated Change*
Thomas Hylland Eriksen

*Identity Destabilised:*
*Living in an Overheated World*
Edited by Thomas Hylland Eriksen and Elisabeth Schober

*Mining Encounters:*
*Extractive Industries in an Overheated World*
Edited by Robert Jan Pijpers and Thomas Hylland Eriksen

# Climate, Capitalism and Communities

## An Anthropology of Environmental Overheating

Edited by
Astrid B. Stensrud and
Thomas Hylland Eriksen

First published 2019 by Pluto Press
345 Archway Road, London N6 5AA

www.plutobooks.com

Copyright © Astrid B. Stensrud and Thomas Hylland Eriksen 2019

The right of the individual contributors to be identified as the authors of this work has been asserted by them in accordance with the Copyright, Designs and Patents Act 1988.

British Library Cataloguing in Publication Data
A catalogue record for this book is available from the British Library

ISBN  978 0 7453 3957 3   Hardback
ISBN  978 0 7453 3956 6   Paperback
ISBN  978 1 7868 0486 0   PDF eBook
ISBN  978 1 7868 0488 4   Kindle eBook
ISBN  978 1 7868 0487 7   EPUB eBook

This book is printed on paper suitable for recycling and made from fully managed and sustained forest sources. Logging, pulping and manufacturing processes are expected to conform to the environmental standards of the country of origin.

Typeset by Stanford DTP Services, Northampton, England

Simultaneously printed in the United Kingdom and United States of America

# Contents

| | | |
|---|---|---|
| List of Figures | | vii |
| Preface | | viii |

1. Introduction: Anthropological Perspectives on Global Economic and Environmental Crises in an Overheated World  
   *Astrid B. Stensrud and Thomas Hylland Eriksen* — 1

2. The Political Economy of the Great Acceleration, or, How I Learned to Stop Worrying and Love the Bomb  
   *Anna Tsing* — 22

3. A Community on the Brink of Extinction? Ecological Crises and Ruined Landscapes in Northwest Greenland  
   *Kirsten Hastrup* — 41

4. Sea Ice, Climate and Resources: The Changing Nature of Hunting Along Greenland's Northwest Coast  
   *Mark Nuttall* — 57

5. Volatility: Understanding Global Capitalism and Climate Change Vulnerability in Mongolia  
   *Andrei Marin* — 76

6. The Dark Side of Progress: The Intersections of Climate Change, Neoliberalism and Modernity in Peru  
   *Astrid B. Stensrud* — 96

7. Puzzling Pieces and Situated Urgencies of Climate Change and Globalisation in the High Arctic: Three Stories from Qaanaaq  
   *Astrid Oberborbeck Andersen and Janne Flora* — 115

8. Counting: Health Emergencies and the Constitution of Extractive Natures in Northern Loreto, Peru  
   *María A. Guzmán-Gallegos* — 133

9. Expansive Capitalism, Climate Change and Global Climate Mitigation Regimes: A Triple Burden on Forest Peoples in the Global South  
   *Harold Wilhite and Cecilia G. Salinas* — 151

10. Climate Change, Oceanic Sovereignties and Maritime
    Economies in the Pacific                                            171
    *Edvard Hviding*

11. Islands of Hope and Despair: Scaling the Collapses and the
    Collapse of Scales                                                  188
    *Frank Sejersen*

12. Using a Glacier Website to Promote Action and Build
    Community: Engaged Anthropology in the Digital Age                  205
    *Ben Orlove, Kerry Milch and Laura Uguccioni*

*Notes on Contributors*    224
*Index*                    228

# Figures

| | | |
|---|---|---|
| 2.1 | The Great Acceleration imagined through J-curves | 23 |
| 2.2 | The home and the world: Hanford signs | 28 |
| 2.3 | 1940s advertisement emphasising the 'great expectations held for DDT' | 31 |
| 5.1 | Mongolia and its ecoregions (study area marked in the rectangle) | 77 |
| 5.2 | Booms and busts of the cashmere price in US$, illustrated by the price paid by factories to primary processors in China (A) and Mongolia (B), and by the price paid by middlemen to herders in Mongolia (C) | 82 |
| 8.1 | The continuous constitution of (river) parts. 'You are beyond [the company's] area of influence. If the river dragged the oil it is not our fault' | 147 |
| 10.1 | The anthropological boundary-making of Oceania | 176 |
| 10.2 | Oceania as 'Big Ocean States' – islands and EEZs | 177 |
| 10.3 | Oceania as a 'Sea of Islands' | 177 |

# Preface

One of the most dynamic growth industries in academic research and publishing is that which deals with anthropogenic climate change and its implications. Every week, there is a new book in the field. Every day, another conference. Every hour, a new research article. This body of work has fanned out to cover the full range of academic disciplines, from geology and hydrology to philosophy and literary studies; sometimes strictly disciplinary, more often with a hint or more of interdisciplinarity. There is general consensus across academia about the reality and severity of humanly induced climate change (although there are inevitably outliers who question it); scholars also by and large agree that it has to be understood from several perspectives. For example, historians may shed light on the present situation by looking at climatic fluctuations in historical times, archaeologists by studying the effects of the great thaw following the last Ice Age; biologists by researching species mobility, extinctions and the effects of changing habitats; cultural studies researchers by looking at the way in which climate change enters popular and high culture, and so on; and at the forefront are the geophysicists and other natural scientists who play a pivotal role in the International Panel on Climate Change (IPCC). Solutions are being proposed, ranging from technological innovations to a shift towards decentralised low-emission societies. The Gaia metaphor, originally suggested to James Lovelock by William Golding, is again being invoked, neologisms like the Anthropocene are coined, and the relative merit of such concepts is being discussed critically, along with the continuous dissemination of new empirical research and the rediscovery, sometimes but not always acknowledged, of older ecological and ecophilosophical thought. Such is the intensity of the academic interest in climate change that it sometimes feels like an overheated domain; crowded, accelerating, frenzied.

The obvious question following from these reflections is why another book about climate change is needed. Fortunately, we can offer several answers. First, the anthropological literature on climate change, focusing as anthropologists do on local life and local understandings in a comparative spirit, is not overrepresented in this field. Since implications of and responses to climate change are diverse, detailed knowledge about many societies, ideally every society worthy of the name, is necessary for a full appreciation of what is sometimes spoken of as 'the human dimension' in climate change. And for change to happen, people have to be in on it, so we might as well find out what they are up to. Second, this book is not just about local responses to climate change, but it also tackles inequality and the centrality of global capitalism as the driving force in change worldwide. Third, this book should ideally be read not only within the context of climate research, or even the anthropology of climate change, but also as a major contribution to the ERC AdvGr project 'Overheating: the three crises of globalisation'.

This is the seventh book to come out of the project, which represents an ambitious bid for a global anthropology. Overheating refers to accelerated global change, and the project aims to describe and analyse localities exposed to fast changes in the domains of the economy, culture and identity, and climate and the environment, but we have also studied mining, labour, waste and boom-and-bust cycles. Former books include an introductory text about overheating (*Overheating*, Thomas Hylland Eriksen 2016), edited volumes about identity (*Identity Destabilised*, Eriksen and Elisabeth Schober 2016), knowledge regimes (*Knowledge and Power*, Eriksen and Schober 2017) and mining (*Mining Encounters*, Robert Jan Pijpers and Eriksen 2018), a monograph from an Australian industrial city (*Boomtown*, Eriksen 2018), and an edited volume, which began as a special issue of *History and Anthropology*, covering several topical areas (*An Overheated World*, Eriksen 2018).

The present book highlights clashing scales in a world where bigger is better and there are few restrictions on international trade and investments; it also takes on the vulnerability/resilience nexus from critical perspectives, documents forms of local adaptations and resistance, and last but not least, it contains original ethnography from many parts of the world. It should nevertheless be conceded that a shortcoming in this book, as in most anthropological work on climate change responses, is the relative lack of ethnographies from those countries, and social groups in them, which are the largest emitters of greenhouse gases and most efficient consumers. There are no chapters from Qatar or Norway here, alas.

Many deserve thanks for bringing this project to fruition. First of all, we would like to thank all the core members of the Overheating group for the inspiring discussions and the encouraging comments received during the planning and writing of this book: Lena Gross, Robert Pijpers, Cathrine Thorleifsson, Elisabeth Schober, Wim van Daele and Henrik Sinding-Larsen. A special thanks goes to the formidable Irene Svarteng for her practical support in the organisation of the workshop in April 2016 that led to the realisation of this book. Overheating was based at the Department of Social Anthropology, University of Oslo, which has given us all the elbowroom and support we could ask for, and we are pleased to see that research related to Overheating continues in new guises at the department.

The title of the workshop was 'Climate Change and Capitalism: Inequality and Justice in an Overheated World'. In addition to the authors in this volume, several others participated as presenters and discussants: Mike Hulme, Susan Crate, Jessica Barnes, Marisol de la Cadena, Lyla Mehta, Karen O'Brien, Marianne E. Lien and Chris Hann. We are grateful to all the participants at the workshop who engaged in the discussions and contributed with valuable comments. But, in particular, we must thank our contributors for their admirable patience and flexibility. We also thank the anonymous reviewers for critical and constructive comments, and the people at Pluto Press for their efforts and encouragement. Finally, a very special thanks goes to the European Research Council for funding this somewhat unconventional research project from 2012 to 2017, under the European Union's Seventh Framework Programme (ERC Grant Agreement no. 295843).

Astrid B. Stensrud<br>
Thomas Hylland Eriksen<br>
Oslo, spring 2019

# 1. Introduction
## Anthropological Perspectives on Global Economic and Environmental Crises in an Overheated World

*Astrid B. Stensrud and Thomas Hylland Eriksen*

The world is overheated, both literally by increasing global temperatures and metaphorically by the accelerated speed of global movements and transactions. In this edited volume, we will explore the interfaces of climate change and capitalism, and discuss how the effects of changes in the climate, weather and environment intersect with the effects of economic policies and social inequality. Social anthropology, as well as other disciplines in the social sciences and humanities, has a lot to contribute to the discussion of such questions of global importance. By examining various interfaces of 'environmental change' and 'economic change', we intend to shed new light on human-environment-capital relationships in a world that is undergoing accelerated change.

The world's populations have become dependent on non-renewable resources, such as coal, oil and gas, and as we are using natural resources and creating $CO_2$ emissions too fast for the planet to regain a viable equilibrium, we are facing a crisis of reproduction. Large-scale natural resource extraction, intensive agriculture and industrial production contribute to environmental damage, the extinction of plants and animals, and global warming. The effects of global warming produce environmental crises on global, regional and local scales, and often lead to enforced changes in livelihoods or displacement owing to droughts, flooding, erratic rainfall or ecosystemic changes. These changes and experiences of crisis happen at an increasingly accelerated speed in a world which can be described as being overheated (Eriksen 2016). The metaphor of overheating does not mainly refer to the physical heating of the planet, but also the speed by which economic, environmental and social changes happen, and the widespread experience that these changes – runaway processes with no regulating governor or thermostat – are getting out of human control, generating a broad range of unintentional long-term consequences with unpredictable, frequently detrimental effects for life on earth.

The chapters in this book discuss how environmental changes and crisis are experienced and dealt with by different groups in various places around the world, and how economic interests, cultural values and social positions shape responses. Perceptions and responses are also scaled in different ways by local communities, national governments or international organisations, and often come into conflict when there are different opinions about whether strategies of adaptation and mitigation are adequate, sufficient, paradoxical or counterproductive. One crucial question is whether we are too dependent on fossil fuels to be able to change structures of production and habits of consumption. Already in 1990, the first Intergovernmental Panel on Climate Change (IPCC) report told us everything we needed to know about dealing with climate change, yet in the subsequent decades we have not just failed to adjust our emissions, but annual emissions were, in 2016, over 60 per cent higher than in 1990 (Anderson and Nevins 2016). After three decades of free market policies and very considerable economic growth, the world is more dependent on fossil fuels than ever. We cannot even begin to imagine the worldwide and long-term ramifications of continued fossil fuel emissions and changes in the atmosphere, but the belief in the need for continued and unhampered economic growth remains uncontested by mainstream politics.

At the Paris climate summit COP21 in December 2015, 195 countries agreed to prevent an increase of global temperature above 2 degrees Celsius compared to preindustrial levels, and even aim to keep the change below 1.5 degrees. A year later, NASA and the National Oceanic and Atmospheric Administration (NOAA) declared 2016 to be the hottest year since modern recordkeeping began in 1880. In 2016, the globally averaged temperatures were 0.99 degrees Celsius warmer than 1951–80 average, and 1.2 degrees Celsius above the average from 1881–1910 (Climate Central 2017; NASA 2017). One would think that the news of the overheated globe would speed up the commitments and actions of governments to enforce policies to reduce $CO_2$ emissions. However, the news was greeted with a deafening silence from the world's biggest emitters. In our home country, Norway, in which the economy and welfare state is dependent on the extraction of oil and gas, the Minister of Petroleum and Energy opened new areas for oil exploration and drilling in a previously protected area in the Arctic in 2016 (*Guardian* 2016), after the Arctic ice edge was redefined by the government in 2015 (Doyle 2015). This move may be an illustration of complacency or even climate change denial, but it may also indicate that the double bind between growth and sustainability runs deep in contemporary modern societies: since growth and prosperity continue to be associated with high consumption of fossil fuels, it is difficult to reconcile the opposing ideals of economic growth and ecological sustainability. This, in the eyes of many, is the key contradiction in today's world (see, for example, Foster et al. 2010; Baer 2012; Eriksen and Schober 2016b; Brightman and Lewis 2017).

In the research project 'Overheating: the three crises of globalisation', a group of anthropologists has examined how different groups of people experience and respond to environmental and economic crises – in addition to crises of identity – around the world (Eriksen 2016; Eriksen and Schober 2016a). Since human beings do not merely 'adapt to' changes in the environment, it must be assumed, at the outset, that people appropriate, engage, interact and negotiate reflexively with their environmental surroundings, as well as with markets, states, corporations and other powerful institutions. This book asks how different groups articulate ideas and claims of blame, responsibility and justice. Given that the causes and effects of climate change overlap with the workings of global capitalism and the continuous striving for economic growth, a central question is how the double bind between material wealth and ecological viability is being negotiated by people with different stakes in the global economy. The chapters that follow offer varied responses and approaches as to how anthropogenic climate change and ecological crisis in different socio-economic and cultural contexts around the world can be understood. Through diverse ethnographic and historical accounts – ranging from fishing in the Pacific and in Greenland, pastoralism in Mongolia and in the Andes, oil spills in the Amazonian rivers, the global politics of forestry and REDD programmes, the military economy in the Cold War, and current climate activism on the Internet – this book discusses key conundrums and tensions in today's world.

It almost goes without saying that there is an urgent need to understand the social and cultural dynamics leading to climate change, and the difficulties of changing political and economic practices accordingly. Since the natural-science approach to climate change does not capture the complexity of human understandings, responses and reactions to it, a complementary literature is needed. Barnes et al. (2013) suggest that anthropology can add to the study of climate change by drawing attention to cultural values and political relations that shape climate-related knowledge creation and that form the basis of responses to continuing environmental changes. Anthropological perspectives and ethnographic methods doubtless have much to contribute to research on global questions by connecting the small scale to global processes. The ethnographic toolkit includes qualitative approaches such as participant observation and narratives, which are tailored to grasp real-life experiences of how the global is constructed and interpreted locally, and how it is scaled in different ways. It enables us to understand that 'the global' is always experienced locally, and more than that, the idea of 'the global' is always produced locally. Global phenomena do not appear from everywhere and/or nowhere, but are generated from specific places and positions. As Kirsten Hastrup (2013: 148) has noted, the local and the global are enfolded in each other, and global climate change 'matters locally wherever it derives from – historically, planetarily or atmospherically'. That people's concerns reach beyond the 'local' requires anthropologists to be aware of the different points of perception and possible scales of attention in ethnographic research (Hastrup 2013).

Participant observation drawing attention to the world as it is perceived and interpreted locally allows anthropologists to see connections that are not obvious at first glance, revealing that 'nature', 'the economy' or 'the environment' may have fundamentally different meanings in different places and for different groups. The report of the American Anthropological Association's Global Climate Change Task Force from 2015 notes that 'anthropology has an important role to play with respect to climate change, and global environmental change, and contributes critical missing pieces to the puzzle' (Fiske et al. 2014: 17). The rationale of this book rests on its ability to provide approaches to how we can understand the 'puzzle' and what these 'critical missing pieces' might look like.

## THE ANTHROPOLOGY OF CLIMATE CHANGE AND GLOBAL CAPITALISM

The significance of economic globalisation has for a great part been overlooked in the anthropology on climate change. Indeed, the very considerable anthropological literature on inequality and global neoliberalism has barely been articulated with that on the environment and climate change – obliquely reflecting a similar schism in politics, where green ideologies sit uneasily with red ones and the two are frequently opposed. In fact, this is an important task because despite the growing evidence of the causes of the current climate and ecological crisis, there still exists a hegemonic belief in the need for continued economic growth on a global scale, and by and large, people worldwide want material security and the right to be consumers. The current capitalist economic system and the accompanying neoliberal ideology of market deregulation are at the core of national, regional and global systemic practices that produce growth in consumption patterns as well as structures of inequality, which make it difficult to achieve a reduction in $CO_2$ emissions and to stabilise global temperature rise in so far as politicians are loyal towards the corporations. This book contends that without looking at the runaway growth dynamics of global capitalism, our understanding of environmental and climate change is incomplete. Hence, there is a need to sharpen the anthropological approach to climate change, incorporating a concern with inequality and the massive impacts of runaway global capitalism and integrating it with ecological perspectives (Foster et al. 2010; Baer 2012; Strauss et al. 2013).

During the present century, the study of climate change has grown massively in anthropology (see, for example, Orlove et al. 2008; Crate and Nuttall 2009, 2016; Hastrup 2009; Barnes and Dove 2015). So far, anthropological research has mostly focused on cultural perceptions and responses to climate change, with a couple of valuable exceptions (Baer 2012; Wilhite 2016). Some scholars have offered critical perspectives on the notions of adaptation, vulnerability and resilience (Crate and Nuttall 2009; Hastrup 2009; Orlove 2009). Others have focused on the psychological and social aspects of climate denial (Norgaard 2011). Others have looked at local health impacts of global warming (Baer and Singer 2009). This is of course very valuable. However, it is time to move beyond the uniqueness and unscalability of the local and begin to address global questions about inequality, capitalism and environmental damage with a basis in ethnography.

In spite of the continued dominance of approaches with a defensive ring – cultural perception and social adaptation – there are currently more anthropologists who are taking the climate/capitalism nexus seriously and who are critical of the notion of 'green capitalism' (Baer 2012; Wilhite 2016). Wilhite (2016) has looked closely at the relationship between our carbon society and consumption habits, arguing that deep reductions in energy use and carbon emissions will not be possible given that our current political economies are driven by the capitalist imperatives of growth, commodification and individualisation. It is therefore necessary to understand the relationship between capitalism and the consumption habits at the level of family and household that are formed in a material world designed and built for high energy use (Wilhite 2016).

In order to survive, capitalism must generate the (artificial) need endlessly to consume a wide array of commodities, even potentially dangerous and lethal ones, such as motor vehicles, which emit pollutants and greenhouse gases (Baer 2012: 41). Calculations based on data from the Global Footprint Network show that if the world's population lived like an average American citizen, we would need four earths to sustain seven billion people (de Chant 2012). Global capitalism and its associated ever-expanding cycle of production and consumption have fostered what has been called a 'toxic culture': 'the unquestioned production of hazardous substances, tolerance for economic blight, dangerous technologies, substandard housing, chronic stress, and exploitative working conditions' (Hofrichter 2000 in Baer 2012: 41). As Andersen and Flora show in their chapter, new patterns of consumption are getting a hold even in remote communities in Greenland. On the other hand, Guzmán-Gallegos demonstrates the deadly consequences of the oil extraction industry, which is the basis of our fossil-fuelled global culture of production, trade and consumption.

We suggest that anthropologists could take inspiration from one of the twentieth-century's most remarkable scholars, namely, Gregory Bateson, who early developed an ecological approach to the world by combining anthropology with philosophy and cybernetics. Bateson (1972) saw the world as a series of systems, where the individual and the ecosystems were inherently dependent upon each other. As early as 1970, he identified three root causes of what he called 'the ecological crisis'. The first he pointed out was technological progress, the second was population increase, and third, he pointed to particular Western cultural values and ideas that concern the exceptional role of mankind in relation to the environment. What Bateson criticised was the idea that humans should strive to control the environment, the strong focus on the individual, the sense of economic determinism, the belief that we live within an infinitely expanding frontier, and that technology will fix things. What Bateson calls a 'healthy ecology' requires ecological flexibility and slow change: 'a single system of environment combined with high human civilisation in which the flexibility of the civilisation shall match that of the environment to create an ongoing complex system, open-ended for slow change' (Bateson 1972: 502).

However, after the Great Acceleration that has taken place in the past half century, our ecological footprint is beyond the point of no return (see also Tsing's chapter in this volume). According to IPCC (2014), continued emissions of greenhouse gases will increase the likelihood of severe, pervasive and irreversible impacts for people and ecosystems. Even if anthropogenic emissions are stopped, many climate change impacts will continue for centuries (IPCC 2014).

## THE ANTHROPOCENE, THE GREAT ACCELERATION AND (UN)SUSTAINABILITY

As suggested, there is no disagreement in the scientific community regarding the reality of humanly induced climate change (IPCC 2014). In the recognition that the human imprint is all over the planet – on the land, in the oceans, in the atmosphere – the term 'Anthropocene' was famously introduced around the turn of the millennium in order to define a current geological epoch replacing the Holocene epoch (Crutzen and Stoermer 2000). What the term makes abundantly clear is that global climate change is a direct consequence of the increased use of fossil fuels and emissions of carbon dioxide, which has skyrocketed since the early nineteenth century and especially after the Second World War. By the 1890s, half of global energy use came in the form of fossil fuels, and by 2015 that share had climbed to nearly 80 per cent (McNeill and Engelke 2014: 2). The escalation since 1945 has been named 'the Great Acceleration' (Steffen et al. 2007; see also Tsing's chapter in this volume), echoing Polanyi's famous term 'the Great Transformation', because just as the market economy is embedded in social relations and institutions, the driving forces behind anthropogenic global ecological change are embedded in societies and traditions (McNeill and Engelke 2014). In August 2016, the Working Group on the Anthropocene (WGA) formally designated the Anthropocene as a new geological epoch starting in 1950. The members of WGA are now working towards an agreement regarding the primary signal – or 'golden spike' – that should be used to define the start of the Anthropocene. Zalasiewicz et al. (2017) state that this timing represents the first appearance of a clear synchronous signal of the transformative influence of humans on key physical, chemical and biological processes on a planetary scale. There are many candidates to choose from, including plastic, aluminium and concrete particles, artificial radionuclides, fly ash particles, changes to carbon and nitrogen isotope patterns, and a variety of fossilisable biological remains: many of these signals will leave a permanent record in the earth's strata. According to an informal vote among the members of the WGA in 2016, a clear majority chose radionuclide signals (plutonium, radiocarbon) associated with the 'bomb spike', as these provide arguably the sharpest and most globally widespread signal (Zalasiewicz et al. 2017).

In her chapter in this volume, Anna Tsing offers a thorough analysis of the cultural political economy of the Great Acceleration, focusing on the development of the atom bomb and the spread of radioactivity in the Cold War. She argues that the nuclear bombs that were produced by US military technology not only helped the Allies win the war, but they also helped establish American post-war hegemony. Across the world, nuclear reactors were built by US exports, political agreements and subsidies during the Cold War. Both everyday leaks and unusual collapses have spread radioactivity across land and water. This was, however, naturalised as a form of 'peaceful' wellbeing. The wastelanding of the Bikini Atoll for US nuclear testing in the 1940s and 1950s is a vivid example of the making of sacrifice zones within the project of building a radioactive world for national security (Tsing, this volume). Looking specifically at the High Arctic Thule Region in Northwest Greenland, Kirsten Hastrup shows that the Cold War, as reflecting a kind of geopolitical overheating, implied a profound militarisation of the Arctic, with dire consequences for the people in Thule who for centuries have lived 'on the brink of extinction'. People were forced off their hunting grounds so that the Americans could build an airbase and a city under the ice, which was to be powered by a nuclear reactor and designed to withstand nuclear war. The city was soon abandoned due to the movements in the ice, yet the nuclear waste – in addition to the toxic waste from the Distant Early Warning (DEW) Line radar stations – ruined the hunting grounds.

The concept of the Anthropocene opens up many possibilities for anthropological analyses and discussions about the destructive outcomes of global, extractive modernity, and many anthropologists have already written extensively on the topic (see, for example, Haraway et al. 2016; Hann 2017; Tsing et al. 2017). Bruno Latour (2017) proposes that the concept of the Anthropocene is a gift for anthropology because it recognises that human activity is having a geological impact on the earth, and natural sciences must now take human agency into account. As an anthropologist of science, Latour argued in his book *We Have Never Been Modern* (1993) that the modern constitution is predicated upon the clear separation between nature and society, and between politics and science. By designating our current era as Anthropocene, geologists have made it clear that nature is not really separate from culture after all, and have thus opened an anthropological breach into mainstream natural science (Latour 2017).

The alternative concept 'Capitalocene' (Malm and Hornborg 2014; Haraway 2016; Moore 2016) directs attention towards specific practices and assemblages of capital, power and resources, and how these practices are part of large-scale projects of extraction and production, wealth accumulation, market formations and global trade. Thinking through the Capitalocene implies giving priority to the dynamics of power and capital when explaining the modern world's dependency on fossil fuels, and locating the origins of this modern world with the rise of capitalism (in its mercantilist incarnation) after 1450, with the formations of markets and global trade routes. These were the driving forces leading to large-scale use of coal after the invention of the steam engine, as well as more recent productions of technologies for the extraction of resources and commodification (Haraway 2016; Moore 2016). If the Anthropocene places humanity in its post-Malthusian multitudes – successful, greedy, proliferating, consuming, polluting – centre stage, the concept of the Capitalocene rather places the burden of proof on the economic system which grew out of Renaissance trade, imperialist expansion and plantation economies.

Foster et al. (2010) argue that today's capitalism has remained what it was from the beginning: an enormous engine for the ceaseless accumulation of capital, propelled by the competitive drive of individuals and groups seeking their own self-interest in the form of private gain. All of the leading economists – including Adam Smith, David Ricardo, Karl Marx and John Maynard Keynes – recognise that capitalism is inherently a process of accumulation and growth. Such a system recognises no absolute limits to its own advance, and the race to accumulate is endless. Hence, there is no way to solve our current global crisis as long as capital accumulation is the primary goal of society. Still, practically all mainstream economists continue to praise economic growth, even if they concede that changes may be needed to ensure that this growth is sustainable (Foster et al. 2010: 29).

The fashionable concept of 'sustainable development' has nevertheless been criticised for being an oxymoron. Brightman and Lewis (2017) argue in their edited book *The Anthropology of Sustainability* that the challenge of sustainability demands much more than the protection or preservation of communities or nature reserves, and more than technical fixes or resource limitations. Rather, achieving real sustainability requires the re-imagination and reworking of communities, societies and landscapes, especially those dominated by industrial capitalism, to build a productive symbiosis between humans and the non-human world. This is an ambitious project that challenges the ideology of progress and development (Brightman and Lewis 2017: 2) and which rejects technological solutions based on a shift to sustainable energy. While the dominant approach to a green shift in the West is embedded in a belief in 'green growth' and technological solutions based on science and modernity, there is an emergent global movement mobilising around the slogans post-extractivism and degrowth (Escobar 2016). Latouche and Harpages (2010) argue that to achieve degrowth and a sustainable future, everybody, but especially the well-off, need to radically change their habits, beliefs and mentalities, which would require a radical transformation of dominant imaginaries. This could, for example, mean rejecting the idea of the exceptional human with the right and obligation to conquer an inert 'nature', and to extract and transform this nature into commodities, profit and waste. The problem of feeding a growing global population without high energy use, however, is not sufficiently addressed within this vision (Gomiero 2018). There is no easy way out. All proposed solutions to the environmental crises and the global climate crisis have serious side-effects, and this is a main reason, apart from mere academic interest, that exploring a broad range of responses to the current crisis is a worthwhile exercise.

In their chapter on climate mitigation regimes and their consequences for people living in forests in the global South, Wilhite and Salinas argue that the rhetorical move from capitalism to green capitalism does not radically change the relationship between capital and nature. On the contrary, the practices of global climate mitigation efforts, such as the UN REDD programme, are grounded in the same economic principles that led to the depletion of forests around the world in the first place. Policy designers present the market as the best way to regulate environmental degradation, aiming to make biodiversity conservation compatible with development. However, sustainability and biodiversity are powerful, historically produced discourses that are linked to the modern ideology of progress and development (Wilhite and Salinas, this volume).

Once the discourse on ecological modernisation and the notion of 'green growth' is accepted, Foster et al. (2010) suggest, the environmental problem becomes merely a question of management and markets. In today's world, we are encouraged to see the economic order in which we live in purely technocratic terms as a 'market system' (Foster et al. 2010: 29–30). This was in fact the essence of Karl Polanyi's (1944 [2001]) work more than seven decades ago. Polanyi's critique of capitalism pointed to its disembedding features, 'lifting out' large-scale economic systems of production and distribution from their social foundations. Accordingly, he predicted that people would react against the marketisation and the alienating disembedding of their lives. Expanding on this main current in economic anthropology, we propose to understand the processes leading to the current climatic situation of the planet as one marked by tensions or clashes between scales, where local reproductive needs conflict with large-scale profit-seeking, not only in economic terms, but also ecologically and politically.

## THE PRICE OF NATURE AND LIFE: VALUE, MONEY AND COMMODITIES

More than 70 years ago, Polanyi warned against the dangers of commodifying nature. He contended that 'leaving the fate of soil and people to the market would be tantamount to annihilating them' (Polanyi 1944 [2001]: 137). He argued that the classical economists' definition of land, labour and money as commodities that are produced for sale is a fiction that could ultimately ruin both nature and human lives. Polanyi's argument was based on the empirical definition of a commodity, according to which labour, land and money are not commodities because labour is only another name for a human activity which goes with life itself, land is only another name for nature, and money is merely a token of purchasing power. The problem with the commodity fiction is that it supports the principle according to which no arrangement or behaviour should be allowed to exist that might prevent the actual functioning of the market mechanism (Polanyi 1944 [2001]). Inspired by Polanyi's indictment of classic liberalism, David Harvey has analysed the impact of neoliberal thought as policy. He argues that the fundamental mission of the neoliberal state is 'to optimize conditions for capital accumulation no matter what the consequences for employment or social well-being' (Harvey 2005: 19).

In today's capitalist system of global markets, people, plants, animals and life-ways are converted into resources with economic value (West 2005; Bear et al. 2015; Tsing 2015). It is important to consider *how* these processes of conversion happen. Anthropologists have a long record of writing about value, and conversion and translation of different forms of value, in terms of gift exchange versus commodity exchange (Kopytoff 1986), different spheres of exchange (Bohannan 1959), the meaning of money (Parry and Bloch 1989), and market value versus moral value (Hart et al. 2010). What most anthropological studies of value agree upon is that value – whether use value or exchange value – is not universal but culturally determined (see, for example, Graeber 2001). In her analysis of translation processes between gold miners, development agencies and the Gimi people in Papua New Guinea, Paige West (2005) claims that the Gimi do not only value forests, plants and animals differently from outsiders; they do not necessarily 'value' them at all, because Gimi do not separate themselves from their environment. West argues that environmental translations should not take for granted that people everywhere approach biological diversity as if it were composed of potentially commodified matter to be used rationally and neutrally as a 'resource' provided by nature. Hence, when writing about ecological crisis and climate change, anthropologists should carefully consider how we allow Western concepts and modes of explanation, founded in the institutional differentiation that in many ways *is* the modern world, to dominate translations and analyses. In her book on the commodification of the matsutake mushroom, Tsing (2015) notes that the history of the human concentration of wealth makes both humans and non-humans into resources for investment. This implies the alienation of both people and things, as if the entanglements of living did not matter. 'Through alienation, people and things become mobile assets; they can be removed from their life worlds in distance-defying transport to be exchanged with other assets from other life worlds, elsewhere' (Tsing 2015: 5). Such processes of standardisation and upscaling are essential to contemporary economic globalisation, where the Goliaths continuously overrun the Davids and force them out of business.

According to Alf Hornborg (2016, 2017), however, the fundamental problems of global sustainability may not be inherent in the market principle in itself as much as in the implications of general-purpose money and the globalised scale of the market. General-purpose money makes all values commensurable, regardless of whether they pertain to the reproduction of human organisms, communities, ecosystems or the world-system. It was the exploitation of globalised price differences, particularly regarding land and labour, which provided the conditions for the turn to fossil fuels in eighteenth-century Britain, which in turn inaugurated anthropogenic climate change and the so-called Anthropocene (Hornborg 2017: 293). The main issue is the idea that everything is interchangeable on the same market. The conceptual universe of general-purpose money assumes that everything has a correct price. This idea makes it possible to purchase human time as well as entire ecosystems as market commodities (Hornborg 2016: 62). The imposition of abstract equivalence among incomparable qualities – the very foundation of capitalist social organisation – is as misleading in terms of its implications for social justice as it is in terms of ecological sustainability (Hornborg 2016: 62; see also Guzmán-Gallegos' chapter), and it also inadvertently narrows down the options by assuming universal comparability between incommensurables.

Several chapters in this book discuss the problematic process of converting things, nature and living resources into commodities through the politics of conversion, for example, in carbon trade (Wilhite and Salinas) or quotas of living resources (Nuttall; Andersen and Flora; Hviding). Andersen and Flora show that several animals in Greenland are now seen as what they call 'cash species': in addition to halibut, walrus, seal and narwhal, the endangered polar bears are among the most important, since they provide both meat and added value in fur, ivory, tusk, skull and claws, which can be sold to tourists, visitors or temporary foreign employees. However, the hunting of polar bears is highly restricted, both to ensure a sustainable population of bears but also to avoid international repercussions that might affect the export of other products, since polar bears have become a global emblem of climate change. However, the power of consumers to protect threatened species through global action might affect local communities and livelihoods in unforeseen ways. As Nuttall (this volume) points out, the transition from hunting to fishing has led to a decline of traditional subsistence cultures and, to some extent, the fragmentation of social life built around kinship, social relatedness and community.

Global consumer activism, or 'fair trade', often has few consequences for those on the low end of the production line. As Marin shows in his chapter, making the trade of cashmere – a luxury commodity on the global markets – into 'fair trade' does not make any difference for the individual herders in Mongolia who still have to struggle with climate uncertainty, price volatility and the intermediaries who buy from herders and sell to manufacturers in China. Since the transition to a market economy that started in 1990, the risk inherent in the pastoralist livelihood was transferred from the collective farm and the nationally coordinated agricultural sector to the individual herder. Hence, their vulnerability to climate change was also exacerbated (Marin, this volume).

## CLIMATE FRONTIERS, INEQUALITY AND EXTRACTIVE FRONTIERS

Climate effects are seen as most intense in areas that are 'climate sensitive' – high altitude (Andes, Himalayas), high latitude (Arctic, Antarctic) and near sea level (Pacific islands) – areas that are mainly inhabited by what Susan Crate (2011: 149) calls 'place-based people': human populations that depend directly and daily upon their local environment for their physical, cultural and spiritual sustenance. The frozen Arctic, the highland Andes, the small islands of the Pacific, the forests and rivers in Amazonia and other rainforests around the world can arguably be understood as 'climate frontiers', and they tend to overlap with what Nuttall (this volume) calls 'extractive frontiers' and 'resource spaces' (see also Sejersen, this volume on the Arctic as a 'new economic frontier'). The Amazonia and other rainforests around the world consist of valuable timber, potential land for plantations and oil reserves in the subsoil (Whilhite and Salinas; Guzmán-Gallegos, this volume); the highland Andean areas consist of large mineral reserves, as well as rain-fed possibilities for dams, hydropower and agribusiness (Stensrud, this volume). The Pacific offers the possibility of large-scale fishing; however, some crucial questions remain to be solved regarding the Exclusive Economic Zones (EEZ) of Pacific atoll nations. What the consequences of the diminishing or disappearing national territories will be for the population and for the size of their EEZs are still not clear (Hviding, this volume). The disappearing ice in the Arctic opens up new economic and extractive opportunities such as halibut fishing, mining (Nuttall, this volume), hydropower and aluminium production (Sejersen, this volume). However, as Sejersen shows in his chapter, when the warming climate in Greenland created new economic opportunities, national labour regulations were changed to facilitate foreign investment. In other words, it is not only the environment that is being exploited, but low-paid foreign workers are also exploited through laws of social dumping in order to ensure the circulation and accumulation of capital in a global economy.

What and who counts – and who can be sacrificed? It seems to be generally accepted that some places, ways of life and people must be sacrificed in the name of a greater good, which is continued economic growth and so-called 'development' on a global scale. It is true, as is often said, that not everything that counts can be counted, and not everything that can be counted counts. Yet sometimes, there is indeed not enough counting and measuring, and it is clear that not everybody counts in large-scale modern projects. In the Peruvian Amazonia, native communities' experiences of polluted rivers, dead fish, suffering bodies and death were not taken into consideration by oil companies or state agencies for a long time. It was only when toxic particles in rivers were counted, and high levels of lead in people's blood were measured, that pollution was officially acknowledged and people's own perceptions became relevant (Guzmán-Gallegos, this volume).

All over the world, communities are sacrificed with reference to the greater good, leading to a clash of scales between 'that which is good for Australia' and 'that which is good for Mr Smith, 666 Koala Drive' (see Eriksen 2018 on clashing scales). Seen from the perspective of climate change, it is abundantly clear that 'the good life' (Tsing, this volume), 'progress' (Stensrud, this volume) and high-energy habits of consumption (Wilhite 2016) are possible only because other livelihoods, people, practices and lifeforms are sacrificed (Guzmán-Gallegos; Hviding, this volume). This is akin to David Harvey's concept of accumulation by dispossession (Harvey 2003). The fact that large-scale transformations are allowed to happen, including species extinction and the rapid disappearance of tropical coral reefs, extends Harvey's useful concept by including ecosystemic features and not just human lives.

Following the historian Dipesh Chakrabarty (2009), this book explores how climate change, refracted through global capitalism on an expanding scale, accentuates the production of inequality that is endemic to capitalism, even though the whole climate crisis cannot be reduced to a story of capitalism. In the twentieth century, socialist economies had a far worse ecological record than capitalist ones, leaving vast ecological disaster areas behind when they ceased to exist, capable of expanding their large-scale industrial enterprises without having to bother with the nuisance of public opinion, a free press or elections. The socialist economies of the last century nonetheless had many features in common with contemporary global capitalism, notably a belief in technologically driven progress, the assumption that natural resources were free, and modelling as well as practices where large-scale projects and quantifiable standards were given pride of place, resulting in frictions and clashes between 'the economy' and locals (Scott 1999). Furthermore, our current ecological crisis cannot be isolated as solely a problem of 'global climate change'. The impacts of climate change are socially distributed across human populations, and those who suffer first and worst tend not to be those who caused the problem (Roberts and Parks 2007). Disaster prevention that treats only symptoms and not political and economic structures is doomed to longer-term failure, as Roberts and Parks (2007: 132) argue: 'The root causes of suffering lie in colonial histories and current relations with the global economy that keep nations vulnerable in many senses of the word.' IPCC acknowledges that a primary cause of vulnerability and the increased risk of climate variability and climate change is the high and persistent level of poverty and inequality in most countries: 'the economic inequality translates into inequality in access to water, sanitation, and adequate housing, particularly for the most vulnerable groups, translating into low adaptive capacities to climate change' (IPCC 2014: 1503).

The terms 'uneven geographical developments' (Harvey 2006) and 'ecologically unequal exchange' (Hornborg 2014) describe how financial investments, profits, income, technology and other benefits from economic growth are unevenly distributed among places and populations around the world. Structural inequalities have been reproduced since colonial times: raw materials are extracted from certain parts of the world and processed in other countries where the supply of cheap labour and precarious working conditions are exploited. The start of the Columbian Exchange and the global trade routes that transported humans, gold, silver, cotton, sugar and coffee across the Atlantic, providing the resources for the early development of industrial capitalism in Western Europe, also created the disproportionate distribution of wealth versus human and environmental degradation. Conversely, risk and vulnerability is disproportionately affecting some places and populations that experience the multiple burdens of economic crisis and environmental change, a process that has been called 'double exposure' to economic globalisation and climate change (O'Brien and Leichenko 2000). We argue that there is a need for more ethnographic studies of the impacts of climate change in terrains marked by social inequality, and also more nuanced analysis of how projects of environmental adaptation and mitigation are implemented in realities of inequality and difference.

In the global South, social movements that link environmental issues to issues of livelihood and justice have existed for a long time (Martínez Alier 1992; Guha 2000), and today several environmental movements are struggling for climate justice (see, for example, Hicks and Fabricant 2016). In some cases, communities and individuals have filed lawsuits against energy companies, holding them responsible for the climate change that threatens their homes and livelihoods (Sejersen, this volume; Walker-Crawford 2017).

Today, climate change brings forth a new type of activism, which is necessarily global in scope, and which is organised through new platforms such as web pages and social media. Orlove, Milch and Uguccioni suggest in their chapter that current environmental crises have created new opportunities for engaged anthropology and activism that is both global and community-based. Aiming to support the voices of local and indigenous people in mountain areas that are affected by the loss of glaciers, the website GlacierHub connects information, action and community on a global level, enabling awareness and concern for climate change. Showing that readers seem to be drawn to posts with political content, like those that discuss the impacts of mining activity on glaciers, they suggest that the website builds awareness, in addition to promoting events and participation in marches and climate conferences. Community members, activists and journalists from glacier countries are also invited to write posts, and thus participate in the co-production of knowledge for a global community.

It needs to be pointed out, at the end, that counterhegemonic initiatives and action groups represent a particular view of risk and of globalisation. In Douglas and Wildavsky's *Risk and Culture* (1982), a distinction is drawn between the 'Centre' and the 'Border'. In order to ensure their cohesion, the critics located on the Border are likely to exaggerate risk by assuming that modest probabilities of catastrophe, for example, in connection with emissions from industry, are very real dangers. At the 'Centre', a more sanguine view of modernity's inherent ability to solve its own contradictions (through, for example, technological solutions to environmental problems) is prevalent, and Douglas and Wildavsky also point out that decisions taken at the Centre tend to emphasise profits and efficiency as paramount values, excluding those that cannot easily be measured. Some decision-makers are, moreover, more easily inclined to take risks than others; they can be likened to those who buy lottery tickets in the awareness that the chance of winning is minuscule. In this important sense, the critical perspectives on global capitalism represented through the cases and approaches prevalent in this book represent a particular point of view. However, when it comes to climate, the risks and remedies come across as different from challenges met earlier (*Risk and Culture* was published in the aftermath of the countercultural blossoming in the USA, long before climate change became an important issue). Since the scientific community (which can in most cases be associated with the Centre) is unanimous in its warnings, the question concerns not so much the risks involved as the appropriate courses of action. This is where the contributions to this book may deserve the attention of non-academics, in providing not just recipes for living – which anthropology has done for over a century – but locally grounded ways of responding to climate change and reacting against its causes.

## REFERENCES

Anderson, K. and J. Nevins. 2016. Planting Seeds So Something Bigger Might Emerge: The Paris Agreement and the Fight Against Climate Change. *Socialism and Democracy* 30(2): 209–18.

Baer, H.A. 2012. *Global Capitalism and Climate Change: The Need for an Alternative World System*. Lanham, MD: AltaMira Press.

Baer, H.A. and M. Singer. 2009. *Global Warming and the Political Ecology of Health: Emerging Crises and Systemic Solutions*. Walnut Creek, CA: Left Coast Press.

Barnes, J. and M.R. Dove (eds). 2015. *Climate Cultures: Anthropological Perspectives on Climate Change*. New Haven, CT: Yale University Press.

Barnes, J., M. Dove, M. Lahsen, A. Mathews, P. McElwee, R. McIntosh, F. Moore, J. O'Reilly, B. Orlove, R. Puri, H. Weiss and K. Yager. 2013. Contribution of Anthropology to the Study of Climate Change. *Nature Climate Change* 3: 541–4.

Bateson, G. 1972. *Steps to an Ecology of Mind: Collected Essays in Anthropology, Psychiatry, Evolution and Epistemology*. London: Intertext Books.

Bear, L., K. Ho, A. Tsing and S. Yanagisako. 2015. Gens: A Feminist Manifesto for the Study of Capitalism, Theorising the Contemporary. Available at: https://culanth.org/fieldsights/652-gens-a-feminist-manifesto-for-the-study-of-capitalism (accessed 31 January 2019).

Bohannan, P. 1959. The Impact of Money on an African Subsistence Economy. *The Journal of Economic History* 19(4): 491–503.

Brightman, M. and J. Lewis (eds). 2017. *The Anthropology of Sustainability. Beyond Development and Progress*. New York: Palgrave Macmillan.

Chakrabarty, D. 2009. The Climate of History: Four Theses. *Critical Inquiry* 35: 197–222.

Climate Central. 2017. 2016 Officially Declared Hottest Year on Record, 18 January. Available at: www.climatecentral.org/news/2016-declared-hottest-year-on-record-21070 (accessed 10 January 2019).

Crate, S. 2011. A Political Ecology of 'Water in Mind': Attributing Perceptions in the Era of Global Climate Change. *Weather, Climate, and Society* 3(3): 148–64.

Crate, S.A. and M. Nuttall (eds). 2009. *Anthropology and Climate Change: From Encounters to Actions*. Walnut Creek, CA: Left Coast Press.

Crate, S.A. and M. Nuttall (eds). 2016. *Anthropology and Climate Change: From Actions to Transformations*, second edition. New York: Routledge.

Crutzen, P.J. and E.F. Stoermer. 2000. The 'Anthropocene'. *IGBP Newsletter* 41: 17–18.

de Chant, T. 2012. If the World's Population Lived Like…, 8 August. Available at: https://persquaremile.com/2012/08/08/if-the-worlds-population-lived-like/ (accessed 31 January 2019).

Douglas, M. and A. Wildavsky. 1982. *Risk and Culture: An Essay on the Selection of Technological and Environmental Dangers*. London: University of California Press.

Doyle, A. 2015. Norway Moves Arctic Ice Edge North, Oil Firms on Alert, Reuters, 20 January. Available at: www.reuters.com/article/us-norway-oil-arctic-idUSKBN0KT0HW20150120 (accessed 10 January 2019).

Eriksen, T.H. 2016. *Overheating: An Anthropology of Accelerated Change*. London: Pluto Press.

——— 2018. *Boomtown: Runaway Globalisation on the Queensland Coast*. London: Pluto Press.

Eriksen, T.H. and E. Schober (eds). 2016a. *Identities Destabilised: Living in an Overheated World*. London: Pluto Press.

Eriksen, T.H. and E. Schober. 2016b. Economies of Growth or Ecologies of Survival? *Ethnos* 83(3): 415–22.

Escobar, A. 2016. Thinking-feeling with the Earth: Territorial Struggles and the Ontological Dimension of the Epistemologies of the South. *Revista de Antropología Iberoamericana* 11(1): 11–32.

Fiske, S.J., S.A. Crate, C.L. Crumley, K. Galvin, H. Lazrus, L. Lucero, A. Oliver-Smith, B. Orlove, S. Strauss and R. Wilk. 2014. *Changing the Atmosphere. Anthropology and Climate Change. Final Report of the AAA Global Climate Change Task Force*. Arlington: American Anthropological Association, 137 pp.

Foster, J.B., B. Clark and R. York. 2010. *The Ecological Rift: Capitalism's War on Earth*. New York: Monthly Review Press.

Gomiero, T. 2018. Agriculture and Degrowth: State of the Art and Assessment of Organic and Biotech-based Agriculture from a Degrowth Perspective. *Journal of Cleaner Production* 197: 1823–39.

Graeber, D. 2001. *Toward an Anthropological Theory of Value: The False Coin of Our Dreams*. New York: Palgrave Macmillan.

Guardian. 2016. Arctic Oil Drilling: Outcry as Norway Opens New Areas to Exploration. *Guardian*, 19 May. Available at: www.theguardian.com/world/2016/may/19/norway-arctic-new-oil-drilling-licences (accessed 10 January 2019).

Guha, R. 2000 *Environmentalism: A Global History*. New York: Longman.

Hann, C. 2017. The Anthropocene and Anthropology: Micro and Macro Perspectives. *European Journal of Social Theory* 20(1): 183–96.

Haraway, D.J. 2016. Staying With the Trouble: Anthropocene, Capitalocene, Chthulucene. In J.W. Moore (ed.), *Anthropocene or Capitalocene?: Nature, History, and the Crisis of Capitalism*. Oakland, CA: PM Press, pp. 34–76.

Haraway, D., N. Ishikawa, G. Scott, K. Olwig, A.L. Tsing and N. Bubandt. 2016. Anthropologists are Talking – About the Anthropocene. *Ethnos* 81(3): 535–64.

Hart, K., J.L. Laville and A.D. Cattani. 2010. *The Human Economy: A Citizen's Guide*. Cambridge: Polity Press.

Harvey, D. 2003. *The New Imperialism*. Oxford: Oxford University Press.

—— 2005. *A Brief History of Neoliberalism*. Oxford: Oxford University Press.

—— 2006. *Spaces of Global Capitalism: Towards a Theory of Uneven Geographical Development*. London: Verso.

Hastrup, K. 2009. *The Question of Resilience: Social Responses to Climate Change*. Copenhagen: The Royal Danish Academy of Sciences and Letters.

—— 2013. Scales of Attention in Fieldwork: Global Connections and Local Concerns in the Arctic. *Ethnography* 14(2): 145–64.

Hicks, K. and N. Fabricant. 2016. The Bolivian Climate Justice Movement: Mobilizing Indigeneity in Climate Change Negotiations. *Latin American Perspectives* 43(4): 87–104.

Hornborg, A. 2014. Ecological Economics, Marxism, and Technological Progress: Some Explorations of the Conceptual Foundations of Theories of Ecologically Unequal Exchange. *Ecological Economics* 105: 11–18.

—— 2016. Post-capitalist Ecologies: Energy, 'Value' and Fetishism in the Anthropocene. *Capitalism Nature Socialism* 27(4): 61–76.

—— 2017. Redesigning Money to Curb Globalisation: Can We Domesticate the Root of all Evil? In M. Brightman and J. Lewis (eds), *The Anthropology of Sustainability: Beyond Development and Progress*. New York: Palgrave Macmillan, pp. 291–307.

IPCC. 2014. *Climate Change 2014: Synthesis Report. Contribution of Working Groups I, II and III to the Fifth Assessment Report of the Intergovernmental Panel on Climate Change*. Core Writing Team, R.K. Pachauri and L.A. Meyer (eds). IPCC, Geneva, 151 pp.

Kopytoff, I. 1986. The Cultural Biography of Things: Commoditization as Process. In A. Appadurai (ed.), *The Social Life of Things. Commodities in Cultural Perspective*. Cambridge: Cambridge University Press, pp. 64–91.

Latouche, S. and D. Harpages. 2010. *La Hora del Decrecimiento*. Barcelona: Octaedro.

Latour, B. 1993. *We Have Never Been Modern*, trans. C. Porter. Cambridge, MA: Harvard University Press.

—— 2017. Anthropology at the Time of the Anthropocene: A Personal View of What is to Be Studied. In M. Brightman and J. Lewis (eds). *The Anthropology of Sustainability: Beyond Development and Progress*. New York: Palgrave Macmillan, pp. 35–49.

Malm, A. and A. Hornborg. 2014. A Geology of Mankind? A Critique of the Anthropocene Narrative. *The Anthropocene Review* 1(1): 62–9.

Martínez Alier, J. 1992. *De la Economía Ecológica al Ecologismo Popular*. Barcelona: Icaria.

McNeill, J.R. and P. Engelke. 2014. *The Great Acceleration: An Environmental History of the Anthropocene since 1945*. Cambridge, MA: The Belknap Press of Harvard University Press.

Moore, J.W. 2016. Introduction: Anthropocene or Capitalocene?: Nature, History, and the Crisis of Capitalism. In J.W. Moore (ed.), *Anthropocene or Capitalocene?: Nature, History, and the Crisis of Capitalism*. Oakland, CA: PM Press, pp. 1–11.

NASA. 2017. NASA, NOAA Data Show 2016 Warmest Year on Record Globally, NASA press release 17-006, 18 January. Available at: www.nasa.gov/press-release/nasa-noaa-data-show-2016-warmest-year-on-record-globally (accessed 10 January 2019).

Norgaard, K.M. 2011. *Living in Denial: Climate Change, Emotions and Everyday Life*. Cambridge, MA: MIT Press.

O'Brien, K. and R. Leichenko. 2000. Double Exposure: Assessing the Impacts of Climate Change Within the Context of Economic Globalization. *Global Environmental Change* 10: 221–32.

Orlove, B. 2009. The Past, the Present and Some Possible Futures of Adaptation. In W.N. Adger, I. Lorenzoni and K. O'Brien (eds), *Adapting to Climate Change: Thresholds, Values, Governance*. Cambridge: Cambridge University Press, pp. 131–63.

Orlove, B., E. Wiegandt and B.H. Luckman (eds). 2008. *Darkening Peaks: Glacier Retreat, Science, and Society*. Berkeley, CA: University of California Press.

Parry, J. and M. Bloch (eds). 1989. *Money and the Morality of Exchange*. Cambridge: Cambridge University Press.

Polanyi, K. 2001. *The Great Transformation: The Political and Economic Origins of Our Time*. Boston, MA: Beacon Press. Originally published 1944.

Roberts, J.T. and B.C. Parks. 2007. *A Climate of Injustice: Global Inequality, North-South Politics, and Climate Policy*. Cambridge, MA: MIT Press.

Scott, J.C. 1999. *Seeing Like a State: How Certain Schemes to Improve the Human Condition Have Failed*. New Haven, CT: Yale University Press.

Steffen, W., P.J. Crutzen and J.R. McNeill. 2007. The Anthropocene: Are Human Beings Now Overwhelming the Forces of Nature? *AMBIO* 36(8): 614–21.

Strauss, S., S. Rupp and T. Love. 2013. *Cultures of Energy: Power, Practices, Technologies*. Walnut Creek, CA: Left Coast Press.

Tsing, A. 2015. *The Mushroom at the End of the World: On the Possibility of Life in Capitalist Ruins*. Princeton, NJ: Princeton University Press.

Tsing, A., H. Swanson, E. Gan and N. Bubandt. 2017. *Arts of Living on a Damaged Planet: Ghosts and Monsters of the Anthropocene*. Minneapolis, MN: University of Minnesota Press.

Walker-Crawford, N. 2017. Andean Farmer Demands Climate Justice in Germany. GlacierHub, 2 February 2017. Available at: http://glacierhub.org/2017/02/02/andean-farmer-demands-climate-justice-in-germany/ (accessed 10 January 2019).

West, P. 2005. Translation, Value, and Space: Theorizing an Ethnographic and Engaged Environmental Anthropology. *American Anthropologist* 107(4): 632–42.

Wilhite, H.L. 2016. *The Political Economy of Low Carbon Transformation: Breaking the Habits of Capitalism*. New York: Routledge.

Zalasiewicz, J., C.N. Waters, C.P. Summerhayes, A.P. Wolfe, A.D. Barnosky, A. Cearreta, P. Crutzen, E. Ellis, I.J. Fairchild, A. Gałuszka, P. Haff, I. Hajdas, M.J. Head, J.A. Ivar Do Sul, C. Jeandel, R. Leinfelder, J.R. Mcneill, C. Neal, E. Odada, N. Oreskes, W. Steffen, J. Syvitski, D. Vidas, M. Wagreich and M. Williams. 2017. The Working Group on the Anthropocene: Summary of Evidence and Interim Recommendations. *Anthropocene* 19: 55–60.

# 2. The Political Economy of the Great Acceleration, or, How I Learned to Stop Worrying and Love the Bomb

*Anna Tsing*

On 16 July 1945, the first nuclear bomb was detonated by the United States Army in the New Mexico desert. Anthropogenic radioactivity entered the atmosphere, flew on the winds, and settled back to the ground. In the series of bombs that has followed, the surface of the earth has changed. If someone in the distant future were to look at earth's geological strata, they would see the line where new radioelements begin piling up.[1]

On the basis of this proposed 'golden spike', some geologists and climate scientists have argued that it is the period since 1945 that should properly be called the Anthropocene, that epoch in which humans outdo glaciers in shaping the surface of the earth. In part, this is because this date is also a convenient marker for what climate scientist Will Steffen and his collaborators call the Great Acceleration, the rapid rise in both 'socio-economic trends' and earth system disturbances (Steffen et al. 2015). Many proponents of the 'Great Acceleration = Anthropocene' story argue that the Great Acceleration has confronted us with severe environmental challenges, which may threaten survival on earth for many multicellular organisms, including humans. I agree: we should be scared. But what is this about? Is the Great Acceleration merely the sedimentation of human changes, made worse through speed and scale? Does it show an innocent and unavoidable trade-off between human wellbeing and environmental destruction? In both cases, I think not. This chapter argues that the Great Acceleration can be better explained as a cultural political economy.[2]

One place to begin is the famous J-curves through which we have learned to visualise the Great Acceleration. Since 1950, measurements of many kinds of human activity indicators, from $CO_2$ concentrations to species extinctions, have shot up rapidly, creating the characteristic 'hockey-stick' shape of these curves. See Figure 2.1 for those presented in

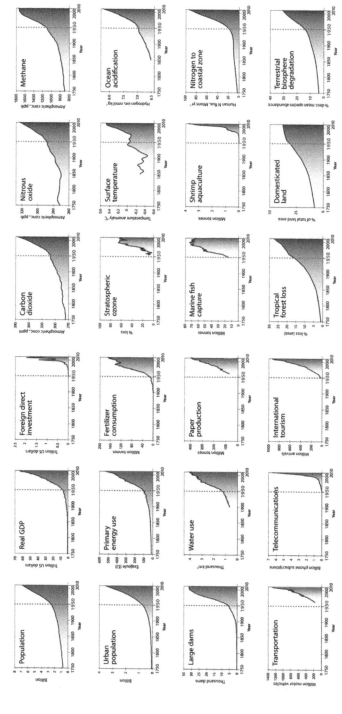

*Figure 2.1* The Great Acceleration imagined through J-curves

Source: Steffen et al. (2015).

the paper by Steffen et al. (2015). Their assembly is an important accomplishment: it narrates an era vividly.

As exciting as these curves are, they also hide things that we need to know. A small story illustrates. In a 2015 geography conference, Steffen showed a J-curve of post-1950 inventions; a cultural geographer in the audience pointed out that inventions – tracked through patents – depend on patent laws to be counted in the first place.[3] The curve tracks the legal changes that make that data emerge. All the curves shown in Figure 2.1 are like that: the data does not come into existence without political, cultural and technical apparatuses. Aggregate data, by which I mean data regarded as a set of neutral and independent measurements, hides those processes through which data emerges. This does not make these curves my enemies; I need them. However, there is a job here for humanists and social scientists: we must re-narrate these curves within the tools we have to know violence, inequality and history. This is not just possible, but also urgent as we respond to threats to planetary liveability. We can tell a better story.

This assignment is much bigger than anything I can handle here; I can, however, open doors for further research. Because this is a door-opening chapter, then, I allow my arguments to be particularly bold. Please do not misconstrue this strategy. Empirical readings inspired the structural arguments rather than the latter dropping from the sky. Still, I begin with them to set the stage for the incomplete investigations that inspired them. These led me to a double-barrelled algorithm for producing the Great Acceleration as Anthropocene:

a. Security = growth. A post-Second World War set of apparatuses aimed at national security worked hard to manufacture growth – and then to naturalise it as human wellbeing.
b. Exclusion = wastelanding. Security required separating privileged insiders and excluded outsiders. Such exclusions were performed by 'wastelanding', that is, by turning landscapes and communities outside the privileged enclave into wastelands.[4]

This algorithm did not emerge from nowhere, and it is also not something timeless such as 'Western civilisation' or even 'modernity'. Instead, I argue, it was formed through a contingent history, as follows:

1. Military technology from the Second World War was used to create post-war growth and security.
2. Cold War competition spread such technologies, with their attendant conceptual worlds, across much of the planet.

3. The naturalisation of these conceptual worlds as human wellbeing meant that late twentieth-century programmes to cut back on Cold War growth projects were only able to chip tentatively, and often in vain, at the edges of those projects.
4. The twenty-first-century 'war on terror' has both revitalised and transformed Cold War projects. The double-barrelled algorithm above has been deployed with ever more force, but not with the same units of inclusion and exclusion as during the Cold War.

The Trinity bomb of the New Mexico desert is a good place to start to tell these stories – and not just because it's a golden spike. The development of the atom bomb begins a Great Acceleration of radioactivity that depends on the historical dynamics I suggest above. Those US bombs helped the Allies win the war – and helped establish American post-war hegemony. Yet their power also framed post-war development. US programmes offered 'Atoms for Peace', an exemplification of the use of military technology to build the Cold War, infused an often extremely violent 'peace' that followed the end of the war.[5] (See #1 above.) US exports, political agreements and subsidies built nuclear reactors across the world. (See #2.) US cooperation with Japan, for example, allowed General Electric to export already questionable reactor designs in the 1970s to erect on Japanese earthquake faultlines, thus creating the conditions for the 2011 Fukushima 'accident'.[6] Both everyday leaks and unusual collapses have spread radioactivity across land and water. Meanwhile, military bombing tests fed the winds – until the tests went underground. The wastelanding of Bikini Atoll for US nuclear testing in the 1940s and 1950s is a vivid example of the creation of sacrifice zones as part of the project of building a radioactive world for national security.[7] When it comes to the spread of radioactivity – certainly an Anthropocene J-curve – my pocket history should be particularly clear. Military technology fuelled the Cold War spread of radioactivity and naturalised it as a form of 'peaceful' wellbeing. Security demanded proliferation – and wastelanding. The double-barrelled algorithm produced threats to liveability.

This is perhaps too easy (and I haven't yet arrived at depleted uranium (Nixon 2011; see #4)). In the rest of this chapter, I try to show that this four-point pocket history is relevant beyond the radioactivity story. For readers not convinced that 'Anthropocene' or 'history' are proper categories for anthropological analysis, please see the conceptual endnote here.[8]

## SECURITY

Itty Abraham's (2009) analysis of the history of 'security' in the United States and Brazil is a useful place to begin broadening the narrative.[9]

There are other places to study security. Yet the concreteness of Abraham's analysis and his willingness to think across cases gives traction to this history, making it possible to ask similar questions elsewhere.

Abraham reminds us of the US National Security Act of 1947, which turned national 'defense' into national 'security' in the United States – while also creating the National Security Council and the CIA. From the first, he argues, national security was haunted by its shadow, insecurity, but what might count as that insecurity to be neutralised was augmented and transformed through a series of events and conjunctures. During the Korean War, US national security became obsessed with the possibility of domestic brainwashing (US prisoners of war had inexplicably refused to be repatriated); after the formation of the RAND corporation, it knew the world through game theory (mathematicians had been enrolled); after Sputnik, with its Cold War competition, the destruction of the world through nuclear retaliation seemed possible.

Meanwhile, in Brazil, army officers who had served under US troops in Second World War Italy went home to initiate an American-inspired War College. When the Brazilian military seized the government in 1964, the officers of the War College were ready to put their concepts of national security into action. *Segurança e desenvolvimento* (security and development) was their slogan for the new Brazil. The murder of Leftists, at home and abroad, was one of their major accomplishments. Abraham argues that even after the end of the dictatorship, in 1985, key features of the security programme were continued. One was *projecão* (expansion): the conquest of the Amazon for national security. The idea that the Amazon was a focus of national insecurity flowed from the history of security as domestic discipline, Abraham argues. No military threats emerged in the Amazon. But the military opening of the Amazon during the dictatorship was continued even more forcefully after 1985. When international advocacy for rainforests and indigenous people emerged in the Rio Earth Summit of 1992, the national security apparatus bit down hard to protect Brazilian 'security'.

Abraham offers scraps of the history from which the algorithms I suggested were constructed. (a) As national security turns inward, towards non-military targets, it encompasses the securitisation of resource zones, such as the Amazon. Resource projects such as dams and mines are sometimes conjured precisely to let security logics play out. (b) Securitisation is the beginning of accumulation by dispossession, David Harvey's term for the displacement of rural and indigenous peoples in the interests of capital (Harvey 2004). Capital is part of the story; so too is security. Exclusions are in the making here, and so is wastelanding.

## THE GOOD LIFE

My next set of clues comes from historian Kate Brown's *Plutopia* (2013). Brown studied two Cold War cities that specialised in the production of plutonium between the 1940s and the end of the Cold War: Hanford, Washington (USA), and Ozersk, in the Russian Urals. The cities mirrored each other in many ways. Both cities were places of pride, family promise and community spirit. Residents felt thrilled to have been chosen to live in each. Each seemed a haven of optimism, safety and beauty. Hanford pioneered the large family houses that came to characterise American suburbs. It recruited healthy families, and because they arrived in such good health, even living with radiation did not seem to hurt them much. Women learned to be happy housewives, while children played at cowboys and Indians. Ozersk families enjoyed the opera, the library and the cinema. Although they had to keep their location secret, for security reasons, they lived enviable lives; they were always happy to get back inside. Hanford families were white, educated and full of hope. Ozersk families became elite citizens. Together, on each side of the Cold War, they pioneered the good life of the second half of the twentieth century.

Brown's research chronicles what it took to create those good lives, and those privileged towns. In Hanford, people of colour built the city, but they did not live in it. In Ozersk, prisoners built the city, but they did not live in it. In Ozersk, radiation polluted the rivers, poisoning rural residents throughout the watershed. In Hanford, radiation went through groundwater, poisoning fish in the rivers and the Native Americans who ate them. Taken together, principles emerge that seem quite important to understanding the good life in the Great Acceleration. The good life is built by enclosing spaces of privilege and imagined safety. Construction workers are not invited inside. Environmental hazards are dumped outside. The irony, of course, is that those hazards sneak back in. But that is not the plan. The plan for a good life is to let others suffer for making your security enclosure and for the hazards you produce.

It seems significant that plutonium-making towns were places that initiated Cold War models for the good life. Plutonium was one of the 'modern' products of the Second World War, deployed afterwards for development as well as war. Its manufacture required the highest national security standards – involving discipline, patriotism and forward-looking attitudes. The good life emerged from that mix. It emerged from the 'friction' (Tsing 2005) between US and Soviet models of modernisation and security. Rubbing up against each other – and thus indirectly shaping each other – a powerful trajectory developed for all those influenced by the Cold War, across the global North and South.

The Great Acceleration was a massive expansion of such models of the good life through development policies on both sides of the Communist/Free World divide.

Again, some structural features: (a) the good life is achieved by exclusions that place hard labour and pollution outside the zone of security and peace; (b) through such means, citizens are enrolled in state projects of environmental destruction; (c) these projects attempt to divide the landscape into zones of safety and sacrifice; (d) national security and Cold War wellbeing depend on such separations (Figure 2.2).

*Figure 2.2* The home and the world: Hanford signs

Source: Silence (top): http://inheritinghanfordblog.com/2013/07/19/toxic-mistrust/; Contamination (bottom): http://hanfordproject.com/.

## FERTILITY

I have already smuggled in two figures of Cold War growth. The first was proliferation, as in nuclear proliferation, a result of the spreading Cold War encompassment of nations and regions across the earth. The second was expansion, which I introduced as a principle of security in discussing the Brazilian military opening of the Amazon. Each of these is essential, I think, to the dreams of progress that fuelled the Great Acceleration. Equally central are projects for the increase of fertility – not only for humans but also for the crops and livestock thought to bring a cornucopia of wellbeing to the earth.[10]

Through fertility, the Great Powers imagined that they could not only inspire their own populations but also take over the world, spreading the Cold War in their favour. Fertility was to be the great gift of Second World War technologies, turned towards what the Cold War called 'peace', that is, zones of securitisation striving for expansion across interstitial spaces of insurgency and counter-insurgency.[11] Crop fertility, the source of animal and human growth, was to be stimulated by two particularly powerful military technologies, which, in the spirit of 'atoms for peace', might be called 'explosives for peace' (fertilisers) and 'poisons for peace' (pesticides and herbicides), respectively.

This history seems key to understanding the Great Acceleration. Both fertilisers and pesticides/herbicides originated from military operations but were re-deployed after the Second World War for Cold War 'peacetime' uses.[12] Both were pushed hard by Cold War development projects until both the global North and the global South became dependent on them. If fertiliser and pesticide/herbicide use (and their attendant ecological effects, from eutrophication to extinction) have increased in J-curves since 1945, it is because of a particular political history of proliferation. Like many bystanders, I had assumed that these chemicals were just signs of progress; reading about them, instead, offered me political programmes.

Every bomber from Oklahoma to Oslo knows that crop fertiliser and explosives are the same.[13] This equivalence motivated the synthesis of ammonia through processes that switched between war and the peace that war engenders. German chemist Fritz Haber did the key scientific work for this synthesis in the early twentieth century; he received his patent in 1911, just two years before German chemistry became integrated into the German mobilisation for the First World War (Smil 2001). By 1914, Haber was working on the production of poison gases to kill enemy soldiers.[14] Ammonia synthesis, in the meantime, was brought into commercial production by techniques developed by Carl

Bosch. During the First World War, the process became industrialised for war munitions. The German war effort had lost access to guano and benefited from synthetic substitutes (Haber 2008: 168). After the war, nitrogen products were used for dyes – but not much for fertiliser. It took the Second World War, and the rapid building of industrial plants on both sides of the war effort, to produce the capacity that could be turned into the new fertiliser industry.

The German war effort in the Second World War benefited not only from the Haber-Bosch process, but also from industrial processes enabled by the same plants, such as the manufacture of synthetic fuels. I.G. Farben, the company that synthesised ammonia, also made synthetic fuel and rubber for the war effort (Smil 2001: 225). A similar process unfolded across the Atlantic, where the US war effort responded by building ten plants for explosive nitrates via the Haber-Bosch process (Philpott 2013). One, Muscle Shoals, had begun to produce munitions during the First World War, but had not been fully developed before the war ended. Between the wars, it had become a fertiliser plant in what became the Tennessee Valley Authority (TVA). During the Second World War,

> TVA used the Muscle Shoals facilities once again for defense industries. More than 60 percent of the phosphorus required for a wide variety of bombs, shells, bullets, and other munitions was produced at Muscle Shoals. TVA also delivered nearly 30,000 tons of anhydrous ammonia, 10,000 tons of ammonium nitrate liquor, and 64,000 tons of ammonium nitrate crystal to the US Ordnance Department. The agency produced more than 200,000 tons of calcium carbide used to manufacture synthetic rubber, employing rehabilitated equipment from World War I. (Ezzell 2009)

After the war, the plant reverted again. 'The TVA became the main research and development center for fertilizer' (Conklin 2009: 111). In 1975 the International Fertilizer Development Center – a Green Revolution proliferator – opened in Muscle Shoals. The pattern evident in other sites is vivid here: wartime technologies were redeployed for post-war development.[15]

The history of pesticides unfolds in a surprisingly parallel fashion. A Swiss chemist, Paul Müller, identified DDT as an insecticide in 1939, but the chemical only achieved its prominence when it entered the arsenal of US Army war chemicals during the Second World War. DDT made the Pacific safe for American troops – by killing typhus-carrying lice and malaria-carrying mosquitoes. During the war, DDT became the new 'wonder insecticide' (Taylor 2007); *Time* magazine made DDT and the atom bomb the two icons of Allied victory (Russell 2001:165). After the war, experts could not wait to use DDT to increase yields in industrial agriculture.[16] DDT also became the most intensely studied chemical in the history of poisons. In what now seems a strange dance with the careful recording of toxicity, every study, even those that recorded danger, made DDT more legitimate, and thus ever more suited for making the good life (Figure 2.3).

*Figure 2.3* 1940s advertisement emphasising the 'great expectations held for DDT'

Source: Killing Salt Chemicals; reproduced in Taylor (2007: 4).

Another class of pesticides, organophosphates, was also developed for wartime use. German chemists first synthesised organophosphates as potential nerve gases for humans, but found they also inhibited nervous system function for other animals, including insects. British and American experts gained access to them through the interrogation of German chemists after the war (Davis 2014: 92).[17] It was American companies, then, that released parathion for commercial use on farms, later joined by EPN, malathion and other relatives. When DDT was banned in the United States (due to the public outcry after Rachel Carson's 1962 *Silent Spring*), these much more immediately toxic chemicals became the standard treatment for industrial agriculture in the United States, and, increasingly, the world.

Herbicides? 'The history of the military weapon Agent Orange,' writes historian David Zierler, 'begins on the eve of World War II, when the demands of total war sparked one scientist's insight that weed killers had military value' (Zierler 2011: 33). Zierler shows how exploratory early research on the chemical killing of plants in the 1930s came to fruition in the US war effort. One leading US scientist thought he could defoliate Japan's forests and even destroy the whole rice crop, starving Japan's population. 2,4-D and 2,4,5-T were tested as early as 1943 in simulated war applications. Hormone herbicides were considered the most viable component of the US programme in chemical and biological warfare. 'Only the rapid ending of the war prevented field trials in an active theater ...' (Zierler 2011: 42). Wartime research opened post-war development: the 'war against weeds'. Americans could still attack Japanese invaders, such as Japanese honeysuckle.

As with the back-and-forth of fertiliser and explosive, herbicide could easily revert to wartime uses. Agent Orange, a mixture of 2,3-D and 2,4,5-T, was deployed in Indochina between 1962 and 1971 by US forces seeking to destroy the vegetation that both underlayed rural subsistence and allowed national liberation forces to hide from imperial air attacks. Great swathes of both mangrove and upland forest were destroyed. Dioxin, a contaminant in the industrial process, was introduced. Both US soldiers and the Indochinese complained about their effects, but without formal redress. Still, a process of public debate ensued, and in 1975, US President Gerald Ford renounced the use of herbicides in war. But what about peace? Why are herbicides necessary for the good life?

American farmers embraced the heady mixture of imperial power and patriotism that characterised the end of the Second World War. Prices were great; after the Marshall Plan began, American farmers could enrich themselves as they fed the world. Fertilisers and pesticides/herbicides were elements of a package, involving machinery and irrigation, which each built on the others.[18] Still, despite their power, these were local cultural commitments. How did parochial US customs proliferate so widely? This question takes me to the Green Revolution, that Cold War initiative in which subsistence crops were bred to grow well only with the chemicals described above. Through the coercive promotion of the Green Revolution, ever-increasing amounts of fertilisers, pesticides and herbicides spread around the world, creating J-curves for the Great Acceleration. Those J-curves show us not only crop increases but also rising tides of eutrophication, harmful algal blooms, dead zones, insect extinctions and many other Anthropocene effects – all results of Cold War agricultural modernisation.

To understand the Green Revolution as a Cold War initiative, one needs a little historical context. Directly after the Second World War, the United States responded to the threat of agrarian revolt, especially in Asia, through land reform programmes. Leftist bureaucrats from the Roosevelt administration were particularly keen to establish programmes in places such as Japan, Taiwan, Korea and the Philippines, where persuading peasants to oppose communism seemed a pressing challenge. After President Eisenhower took control, however, Leftist land reform programmes were abruptly withdrawn, and the United States moved to consolidate alliances with Rightist elites throughout the global South.[19] The rhetoric of rural improvement, however, did not disappear; it was still needed to fight communism. Instead of challenging class privilege, the new programmes moved to a technical register in which improvement consisted of inputs that, in the name of feeding the poor, would yet advantage elites. This was the milieu in which the Green Revolution flourished.

The Green Revolution began with Ford and Rockefeller Foundation attempts to breed a modern, productive wheat strain in Mexico, beginning in 1943. At the heart of the effort was the crossbreeding of a dwarf strain of wheat and a high-producing strain. When the small hybrid plants were drowned in fertiliser – like geese force-fed for fois gras – they responded by reproducing like crazy. In these reproduction machines, almost no energy went into the tiny plant bodies; almost all of it went to the grains. But these force-fed fields produced more than grains: insects and weeds responded to the nutritional bounty by exploding in numbers and vigour.[20] Suddenly pesticides and herbicides were no longer optional; force-fed fields required them.

The wheat-breeding effort was so promising that the Ford-Rockefeller consortium moved on to rice. After the revolution in China, it was rice-eaters who seemed most likely to harbour communist sympathies; the new rice strains, some experts imagined, might fight the Cold War for them. Because this was supposed to be an effort for and from the global South, the International Rice Research Institute (IRRI) was founded in 1960 in the Philippines, where it continued to dialogue with Mexican improved wheat – and then maize – institutionalised as the International Maize and Wheat Improvement Center. Support and success rolled in, and soon the consortium was ready to join with the World Bank and the Food and Agriculture Organization (FAO) to tackle Africa. The Consultative Group on International Agricultural Research (CGIAR) formed in 1971 to draw these efforts together across the 'food-deficit' countries of the world. Working with Cold War dictators across the global

South, Green Revolution technologies were put into place in the great food-producing regions of the world, spreading agricultural chemicals.

Almost all the social scientists who went to study what happened when the Green Revolution came to particular places were appalled at what they found. Green Revolution technologies required capital for seeds and chemicals. They encouraged land consolidation and mechanisation, driving out smallholders and eliminating the occupations of the poor. In Java, the Green Revolution took away women's traditional work, forcing them into new dependence (Stoler 1977). In Malaysia, the new rice technology destroyed peasant 'moral economies', creating new forms of greed in which poverty became more grinding (Scott 1987). In the Philippines, official backing forced peasants into new forms of subterfuge (Nazarea-Sandoval 1995). New pests and herbicide-resistant weeds spread as a result of the overuse of poisons; pollinators disappeared; eutrophication and harmful algal blooms destroyed fishing livelihoods as an effect of the massive use of fertiliser. Yet it seems that the Green Revolution has shown considerable resilience, despite the many criticisms directed at it. Many small reforms have been introduced, especially since the end of the Cold War: gender and development; basic needs; alternative technologies; TEK. But the structural features of the programme have remained in place, and particularly the 'common sense' that the good life can be achieved through technologies of growth. How does this work?

Several of my sources on fertilisers and pesticides examine the scientific testing that has characterised their development.[21] What stands out to me is the strange fact that – whatever testing showed – it never interfered with the common sense of Cold War securitisation through growth. Testing may establish more effective use, but it never seems to inspire calls to ban a poison or stop a form of pollution. Testing becomes part of the algorithm of a better life through security's externalisations; testing assumes that polluting elements can be removed 'somewhere else'. In contrast, when public campaigns have protested at the use of a chemical, it is always because some set of critics has stepped outside to look at the product from such a 'somewhere else'.

I mentioned, for example, the huge amount of testing of DDT in its early years; none of it led to criticisms of the poison, even as its toxic effects became better known. Labelling or more careful use were suggested by the testers, who continued to imagine protected sites of non-exposure (Davis 2014: 120). In contrast, Rachel Carson (1962) began with songbirds, a set of victims outside of the nexus of agricultural modernisation, with its self-enclosed thinking. Similarly, fertiliser experts suggested more careful application and more frequent soil testing – as if this would stop eutrophication, dead zones, harmful algal blooms and the extinction of organisms that do not thrive on force-fed nutrients. If the latter processes still seem irrelevant, this is evidence of what I called the double-barrelled algorithm of the Great Acceleration. The good life is achieved through imagining security zones not affected by the externalisation of cheap labour and unregulated pollution. In this logic, there is always somewhere else to put the trash.

CGIAR has sponsored a report on its formation (Özgediz 2012); I read it to inform my history. One of the most surprising features of *CGIAR at Forty* is that it is chock full of beautiful photographs of rural and indigenous people engaged in traditional livelihoods: an African pastoralist carrying a lamb; a Philippine girl by her bamboo-and-thatch house; or colourfully dressed Andean women and their potatoes. There is some kind of 'imperialist nostalgia' here, as Renato Rosaldo (1989) uses the term; these images evoke tender feelings towards the very worlds the programme has tried so hard to destroy. In this contradiction, I find something important about the persistence of the Green Revolution after the end of the Cold War, a key element for understanding the Great Acceleration. Through these images, the report seems to be saying that the Green Revolution supports cultural diversity. Fertility is a gift for anyone, whatever their status, the images tell us. This is how Cold War common sense about the good life has come to work, even as the Cold War has disappeared.

Abundance, we learn to understand, does not block diversity or equality; it is a neutral aid to any cultural goal we choose to pursue. The good life is possible for all – as long as we can learn to look the other way as we secure our spaces against cheap labour and killing pollution. Even indigenous people, the photographs tell us, can find some place to dump the refuse of the Second World War and post-war American hegemony.

## CONTINUOUS GROWTH IS CONTINUOUS WASTELANDING

Since the end of the Second World War, both earth systems and human social indicators have shot upward in a set of mouth-dropping J-curves. From population to pollution to $CO_2$ levels, all indicators of human activity have flown off the charts. This chapter offers some opening analysis of how to understand the dangerous dynamic called 'growth' without the pretence that aggregate numbers speak for themselves. Non-sustainability developed not as an unintentional effect, I have argued, but as an active strategy of states, corporations and citizen groups on both sides of the Communist/Free World line. Non-sustainability was an ideology for each; we called it 'growth'. Population promotion, chemical pollution, the reckless use of fossil fuels, eutrophication and the spread of radioactivity were active programmes. Now that commentators have pointed out the ecological challenges, everyone wants to deny that they were caused deliberately; in such denials, we imagine such effects as the human condition – rather than as results of a political culture. Instead of naturalising out-of-control human disturbance of planetary ecosystems, I suggest that we follow the Cold War race to modernisation – and its contemporary continuations in the war on terror. This chapter is just a starting point; it opens a door. Thus, I return to my opening algorithms: security = growth; exclusion = wastelanding. If we want to know how our current environmental mess got quite so big so quickly, this cultural political economy is a place to begin.

## NOTES

1. This chapter originates from a class presentation I gave to explain why I had assigned Joe Masco's *The Theater of Operations* (2014; about continuing Cold War effects) and Tracy Voyles' *Wastelanding* (2015; about Navajo country uranium mining) in the same week. I found the conjuncture a surprisingly good way to introduce a post-1945 Anthropocene; afterwards, I itched to explore how far this road might take me into the Great Acceleration. This chapter offers a few more steps, but it is still an interpretation of Masco's and Voyles' work, to which I am indebted. While I have looked for more directly focused histories of the Cold War's environmental effects, those I have found so far interpret both 'environment' and 'Cold War' too narrowly for the project of this chapter (for example, McNeill and Unger 2010). I am grateful to Zachary Caple and our seminar in 'Critical Landscape Ecology' for initiating these thoughts. Zahirah Suhami and Kali Rubaii kindly commented on a draft. My title references Stanley Kubrick's *Dr Strangelove*. www.imdb.com/title/tt0057012/ (accessed 30 March 2019). The conference version of this chapter also tackled the post-1945 rise in human population. www.sv.uio.no/sai/english/research/projects/overheating/seminars/climate-change-and-capitalism%3A-inequality-and-just.html (accessed 30 March 2019).
2. There is something particular about post-1945 modernisation and development, yet many longer histories come together to inform the Great Acceleration. To put it in context, consider other proposed dates for the start of the Anthropocene, including domestication 8,000 years ago (Ruddiman 2003); the European conquest of the New World with its attendant human genocide and ecological transfers (Lewis and Maslin 2015); and the industrial revolution (Crutzen 2006).
3. Discussion during Chair's Opening Plenary, Royal Geographical Society Annual International Conference, Exeter, UK, September 2015.
4. See Voyles (2015) for the term 'wastelanding'.

5. 'Atoms for Peace' was the title of a speech by US President Eisenhower in 1953. www.iaea.org/about/history/atoms-for-peace-speech (accessed 30 March 2019).
6. www.globalresearch.ca/fukushima-general-electric-knew-its-nuclear-reactor-design-was-unsafe-so-why-isnt-ge-getting-any-heat-for-fukushima/5361300 (accessed 30 March 2019).
7. See, for example, Johnston and Barker (2008).
8. *Anthropocene* is a disputed term; indeed, this is what makes it interesting to me (Swanson et al. 2015). It offers a new opening to the conversation between humanists and scientists on the nature of nature. In contrast to stereotypes of scientific rejection of politics, here is an occasion in which natural scientists are mobilising a rather radical agenda. That does not mean we have to accept the terms in which it comes to us. Critical social scientists can turn the discussion around through our intellectual and political insights. My first step towards making Anthropocene responsive to questions of political economy and culture is to bring it into *history* – yet here is another disputed term. By the word 'history', I do not mean national modernisation narratives. Instead, I refer to time-infused tracks and traces, human and not human, and the stories told about them, secular and otherwise (Tsing 2015). Anthropocene histories are radical criticism to the extent they allow us to denaturalise hegemonic practices of human disturbance. Rather than imagining this Anthropocene as a single, global trajectory of human progress and environmental decline, I argue for the importance of many uneven, unequal and far-from-preordained histories, which interact through contingency and conjuncture (Tsing et al. 2019). Rather than put up with the common sense of 'trade-offs' between human wellbeing and environmental destruction, I explore the histories that have conjured such strange forms of thoughtless common sense.
9. See also Abraham (1998), which discusses post-Second World War security in relation to India's nuclear industry.
10. The question of human fertility proved too difficult for this chapter, but see the pioneering essays in Clarke and Haraway (2018).
11. Mamdani's (2005) discussion of the importance of proxy wars in the global South as a Cold War strategy is a useful introduction to this geography.
12. The Second World War is not the only place to begin. The First World War transformed the chemical industry, making it bloom, and established the ties between the use of chemicals in war and peace (Russell 2001: ch. 2). The Second World War brought those earlier dreams to fruition.
13. I am thinking here of Timothy McVeigh and Anders Breivik.
14. In 1915, Haber's wife shot herself, some say because she hated her husband's research; Haber won the Nobel Prize in 1919, by which time he was commonly criticised as 'the inventor of gas warfare'. Ironically, Haber, of Jewish descent, was forced to resign his position in 1933, after Hitler's rise to power; he went into exile in the UK (Smil 2001; Haber 2008).
15. '[A]s a result [of wartime production of ammonia and the conversion of wartime plants to fertilizer production], there was a rapid increase in fer-

tilizer consumption. The advantages of fertilizers were emphasized, and production capacity increased in leaps and bounds. From 1940 to 1950 the number of ammonia plants doubled; then from 1950 to 1960 the number doubled again' (Maxwell 2004: 25).

16. War metaphors spurred civilian use, just as insect metaphors had facilitated wartime plans to exterminate the enemy (Russell 2001: 110, 120). Chemical warfare against insects became best practice. Daniel (2005: 13) quotes a US Science Service article published just after the Second World War: 'Only a very poor farmer does not use a variety of chemicals to help him raise his crops these days.' He notes the importance of chemical companies such as Geigy, Du Pont and Hercules in making the transition from wartime to domestic production in the United States.

17. The US government expropriated war secrets, allowing US companies access to intellectual property claimed elsewhere (Russell 2001:146). Meanwhile, organophosphates strengthened material and metaphorical ties between war and agriculture; as one contemporary put it: 'the line between gas warfare and battling insect pests has become very thin' (Mathews 1950: 379 quoted in Russell 2001:172). As in fertiliser production, chemical plants were redirected from weapons production to pesticide production (Russell 2001: 174).

18. See www.livinghistoryfarm.org/ (accessed 30 March 2019) for personal testimony about the syncretic effects of these new introductions.

19. Putzel (1992) offers an introduction to this history.

20. Gan (2017) explains how the Green Revolution in the Philippines made the brown planthopper a new kind of rice pest.

21. See particularly Davis (2014) and Smil (2001).

## REFERENCES

Abraham, I. 1998. *The Making of the Indian Atomic Bomb*. London: Zed Books.
—— 2009. National Security/Segurança Nacional. In C. Gluck and A. Tsing (eds), *Words in Motion: Toward a Global Lexicon*. Durham, NC: Duke University Press, pp. 21–39.
Brown, K. 2013. *Plutopia: Nuclear Families, Atomic Cities, and the Great Soviet and American Plutonium Disasters*. Oxford: Oxford University Press.
Carson, R. 1962. *Silent Spring*. New York: Houghton Mifflin.
Clarke, A. and D. Haraway. 2018. *Making Kin Not Population*. Chicago, IL: Prickly Paradigm Press.
Conklin, P. 2009. *A Revolution Down on the Farm: The Transformation of American Agriculture since 1929*. Lexington, KY: University Press of Kentucky.
Crutzen, P. 2006. The Anthropocene. In E. Ehlers and T. Kraft (eds), *Earth System Science in the Anthropocene*. Berlin: Springer, pp. 13–18.
Daniel, P. 2005. *Toxic Drift: Pesticides and Health in the Post-World War II South*. Baton Rouge: Louisiana State University Press.
Davis, F. 2014. *Banned: A History of Pesticides and the Science of Toxicology*. New Haven, CT: Yale University Press.

Ezzell, P.B. 2009. Tennessee Valley Authority in Alabama (TVA). In *Encyclopedia of Alabama*. Available at: www.encyclopediaofalabama.org/article/h-2380 (accessed 30 March 2019).

Gan, E. 2017. An Unintended Race: Miracle Rice and the Green Revolution. *Environmental Philosophy* 14: 61–81.

Haber, T. 2008. *The Alchemy of Air: A Jewish Genius, a Doomed Tycoon, and the Scientific Discovery that Fed the World but Fueled the Rise of Hitler*. New York: Broadway Books.

Harvey, D. 2004. The New Imperialism: Accumulation by Dispossession. *Socialist Register* 40: 63–87.

Johnston, B.R. and H. Barker. 2008. *Consequential Damages of Nuclear War: The Rongelap Report*. Walnut Creek, CA: Left Coast Press.

Lewis, S. and M. Maslin. 2015. Defining the Anthropocene. *Nature* 519: 171–81.

Mamdani, M. 2005. *Good Muslim, Bad Muslim: America, the Cold War, and the Roots of Terror*. New York: Harmony.

Masco, J. 2014. *The Theater of Operations*. Durham, NC: Duke University Press.

Mathews, S. 1950. Gas Warfare on the Farm. *Science News Letter*, 7 December: 378–9.

Maxwell, G. 2004. *Synthetic Nitrogen Products*. Berlin: Springer Science and Business Media.

McNeill, J.R. and C. Unger. 2010. *Environmental Histories of the Cold War*. Cambridge: Cambridge University Press.

Nazarea-Sandoval, V.D. 1995. *Local Knowledge and Agricultural Decision Making in the Philippines: Class, Gender and Resistance*. Ithaca, NY: Cornell University Press.

Nixon, R. 2011. *Slow Violence and the Environmentalism of the Poor*. Cambridge, MA: Harvard University Press.

Özgediz, S. 2012. *The CGIAR at Forty: Institutional Evolution of the World's Premier Agricultural Network*. Montpellier: CGIAR.

Philpott, T. 2013. A Brief History of our Deadly Addiction to Fertilizer. *Mother Jones*. Available at: www.motherjones.com/tom-philpott/2013/04/history-nitrogen-fertilizer-ammonium-nitrate (accessed 30 March 2019).

Putzel, J. 1992. *A Captive Land: The Politics of Agrarian Reform in the Philippines*. New York: Monthly Review Press.

Rosaldo, R. 1989. Imperialist Nostalgia. *Representations* 26: 107–22.

Ruddiman, W. 2003. The Anthropogenic Greenhouse Era Began Thousands of Years ago. *Climate Change* 61: 261–93.

Russell, E. 2001. *War and Nature: Fighting Humans and Insects with Chemicals from World War I to Silent Spring*. Cambridge: Cambridge University Press.

Scott, J. 1987. *Weapons of the Weak*. New Haven, CT: Yale University Press.

Smil, V. 2001. *Enriching the Earth: Fritz Haber, Carl Bosch, and the Transformation of World Food Production*. Cambridge, MA: MIT Press.

Steffen, W., W. Broadgate, L. Deutsch, O. Gaffney and C. Ludwig. 2015. The Trajectory of the Anthropocene: The Great Acceleration. *The Anthropocene Review* 2(1): 81–98.

Stoler, A. 1977. Class Structure and Female Autonomy. *Signs* 3(1): 74–89.

Swanson, H., N. Bubandt and A. Tsing, 2015. Less than One and More than Many, Anthropocene as Science Fiction and Scholarship-in-the-Making. *Environment and Society* 6(1): 149–66.

Taylor, E. 2007. Pesticide Development: A Brief Look at the History. *Southern Regional Extension Forestry*, SREF-FM-010. Available at: https://sref.info/resources/publications/pesticide-development---a-brief-look-at-the-history (accessed 30 March 2019).

Tsing, A.L. 2005. *Friction: An Ethnography of Global Connection*. Princeton, NJ: Princeton University Press.

—— 2015. *The Mushroom at the End of the World: On the Possibility of Life in Capitalist Ruins*. Princeton, NJ: Princeton University Press.

Tsing, A.L., A.S. Mathews and N. Bubandt. 2019. Patchy Anthropocene: Landscape Structure, Multispecies History, and the Retooling of Anthropology. *Current Anthropology* 60: Supplement 10.

Voyles, T. 2015. *Wastelanding: Legacies of Uranium Mining in Navajo Country*. Minneapolis, MN: University of Minnesota Press.

Zierler, D. 2011. *The Invention of Ecocide: Agent Orange, Vietnam, and the Scientists Who Changed the Way We Think About the Environment*. Athens, GA: University of Georgia Press.

# 3. A Community on the Brink of Extinction?
## Ecological Crises and Ruined Landscapes in Northwest Greenland

*Kirsten Hastrup*

This chapter is based on historical research and ethnographic fieldwork in the High Arctic Thule Region in Northwest Greenland over many years. This region is home to some 700 people, known as Inughuit or the Thule people, living in the northernmost community on the globe, marginal even within Greenland, and currently greatly challenged by global warming. The massive changes to the ice cover affect the hunt, and their particular mode of living is on the wane. Clearly, and despite geographical marginalisation, their circumstances reflect accelerating global trends of climate, geopolitics and wildlife protection.

On closer inspection, this High Arctic community has faced environmental challenges since the nineteenth century, when the people became known to outsiders. Since then, they have been reported as living on the edge of the humanly possible, and often portrayed as doomed. Yet, historical sources from the past 200 years show how they have survived against all odds during previous periods of environmental insecurity. Their technological ingenuity is legendary, however, during long periods of environmental and geopolitical upheaval their resilience has been strained to the limit. Relating my own observations and looking back into the history of this community, I hope to show what it implies to live on the brink of extinction.

This is how they have often been cast, and have often felt, as we can read in the historical sources, and as we can detect in conversation. There have been several instances of demographic near-extinction, of dramatic loss of vital technologies, of famine and epidemics, of colonial encroaches that actually turned out to be beneficial, and of a military presence that ruined their landscape. Currently, it is climate change and geopolitics that seem to undermine their community, as it accelerates in the North and creates a *New Arctic*, 'referring to a recent era in circumpolar history set in motion by an unparalleled confluence of political and natural phenomena' (Doel et al. 2014: 2). Meanwhile, the view of the Arctic landscape has changed from being powerful and even hostile, to being fragile and in need of protection; this challenges the scientific community in new ways. Perhaps paradoxically, the emergence of the New Arctic calls for penetrating historical analyses; without understanding the long-term processes of colonisation, resource exploitation and military logic, we may not be able to see the immensity of what is happening now (Doel et al. 2014: 6). Equally important, in making such analysis we should, as Laura Ann Stoler has it, 'refocus our historical lens on distinctions between what is residual and tenacious, what is dominant but hard to see, and not least what is emergent in today's imperial formations – and critically resurgent in responses to them' (Stoler 2008: 211).

The New Arctic is not only fraught with ancient geopolitics but also with new ecological nostalgias that reshape local communities as heritage to be protected. It is this process that I wish to discuss here through a close-up on three historical periods. First, the nineteenth-century discovery and the community's enrolment into a global order of trade and disease; second, their being taken hostage to international Cold War politics in the 1950s and beyond; third, their exposure to late-industrial contamination of the environment and to countermeasures of protection that potentially reduce their life to heritage, framing a dynamic community as a still picture.

## OPENING WATERS: THE GLOBAL MARKET OF DISCOVERY AND DISEASE

In the nineteenth century, the Little Ice Age that had held the High North in its grip for centuries began to abate. This meant that the isolation of the Northwest Greenlanders was broken after some three centuries of being cut off from their southerly neighbours, all of them having once entered Greenland across the narrow Smith Sound between America and Northwest Greenland from 1250 onwards. Whatever the Danes, gradually building their small colonies on the (south)western Greenlandic coast since 1721 onwards, thought about Greenland, it never included the lands in the far Northwest and its inhabitants. They remained unknown, even if tales about wild people in the North may have circulated.

In the Arctic, European travel and exploration always implied an act of stretching the limits of the world. This was at least partly a consequence of a particular European way of thinking about maps and of relying upon ancient metaphors, while cultivating the 'planetary consciousness' of the Enlightenment (Pratt 1992: 15ff). The far North had to wait longer than most other regions to become part of this consciousness; it was largely inaccessible and the imagination had to be stretched to accommodate it. One pertinent image in European representations of the far North was the ancient myth of *Ultima Thule*, situated on the edge of the horizon (Hastrup 2007). When the ice-packed waters in the North began opening up in the nineteenth century, European interest in the region was awakened, including an old dream of finding a Northwest Passage to ease trade with the Far East. Apart from the receding ice, there was also a huge British fleet – having become purposeless after the Napoleonic wars – pushing in the same direction. This geopolitical contingency accounted for much of the Arctic activity that took off and for the European discovery of a community in the High North.

The earliest documented encounter took place in August 1818, with the Scotsman John Ross' voyage of discovery on behalf of the British Admiralty, whose explicit aim was to find the Northwest Passage. In this Ross did not succeed, yet he discovered uninhabited dwellings and a small group of people. His account of the encounter with the unknown tribe is fascinating; as Ross wrote: 'They exist in a corner of the world by far the most secluded which has yet been discovered' and that they have 'until the moment of our arrival, believed themselves to be the only inhabitants of the universe, and that all the rest was a mass of ice' (Ross 1819: 123–4). Actually, we cannot truly know what they thought of themselves; Ross only met a dozen men for a single day, and concluded – as other explorers would do in the equally enigmatic Pacific – that such distant and 'unknown' (to the explorers) people hardly knew themselves (Bravo 1998).

In the wake of Ross' pioneering voyage across the Melville Bay, whalers and more explorers soon followed. The region opened up to new forms of local exchange and long-distance trade that brought the Northwest Greenlanders into the global market economy – if in a small way. Each summer, groups of Arctic Highlanders (as Ross had dubbed them) would congregate at Cape York, a promontory in the southern part of the region, hoping for opportunities for barter. The sailors were ready to comply and provided new materials and foreign goods in exchange for local furs; this was of major importance in the region after centuries of isolation. With the help of locals, Elisha K. Kane, who sojourned in the region from 1853 to 1855, counted about 140 persons distributed along the ragged coastline that covered some 1,000 km, on which they lived. At this point in time, the people were suffering from famine and saw themselves as more or less doomed. Still, their sense of community was remarkable, as was their keeping track of each other along the coast (Kane 1856, II: 211).

Meanwhile, there was a growing sense of an outer world offering new possibilities in an insufficient subsistence economy. However, ten years after Kane's report, the population was little more than 100 people according to Isaac I. Hayes, who had been with Kane on the expedition mentioned above, and now came back on his own (Hayes 1866: 386). Hayes reported on abandoned settlements since the last expedition and related how one of his companions had become very sad when queried about the fortune of his own people and had said: 'Alas . . . we will soon be all gone.' Hayes responded that he would come back, to which his companion replied: 'Come back soon . . . or there will be none here to welcome you.' This made Hayes reflect further and to write:

> To contemplate the destiny of this little tribe is indeed painful. There is much in this rude people deserving of admiration. Their brave and courageous struggles for a bare subsistence, against what would seem to us the most disheartening obstacles, often being wholly without food for days together and never obtaining it without encountering danger, makes their hold of life very precarious. The sea is their only harvest field; and, having no boats in which to pursue the game, they have only to await the turning tide or changing season to open cracks, along which they wander, seeking the seal and walrus which come there to breathe. The uncertain fortunes of the hunt often lead them in the winter time to shelter in rude hovels of snow, and in summer, the migrating water-fowl come to substitute the seal and walrus, which, when the ice fields have floated off, they can rarely reach. (Hayes 1866: 386)

Here, there is a clear sense of pending extinction. Part of the price paid by the people, eagerly awaiting the arrival of foreigners and goods since the first contact with John Ross, was to be exposed to epidemics that proved fatal to many in the non-resistant population (Gilberg 1976; Hastrup 2018). This may account for the rapid decline noted by Hayes, even if it cannot be verified, except through circumstantial evidence.

In addition to epidemics, the harshness of the High Arctic environment also contributed to the difficulties in maintaining the fragile economy. This was not only because of a decline in the living resources, but also for want of driftwood and thus a degradation of their ancient technologies of transport and hunting; they had no longer any boats, for instance, as Hayes noted. Previously, a trickle of driftwood had reached their shores by the tail-end of a sea-current going up along the west coast of Greenland, but this had failed for a long time, and left their hunting gear remarkably reduced (Hayes 1866: 414). Hayes muses: 'It is sad to reflect upon the future of these strange people; and yet they contemplate a fate which they view as inevitable, with an air of indifference difficult to comprehend' (Hayes 1866: 415).

Ross, Kane and Hayes all commented on the low level of technology, owing to the absence of wood, leaving them to construct the vital dog sledges entirely out of walrus and narwhal bone, meticulously pieced together by sealskin string – and amazingly well functioning. The full extent of the technological loss was evident to a group of Baffin Landers, who migrated across the narrow strait between Ellesmere Island in Canada and Northwest Greenland in the late 1860s. They amounted to a total of 15 people, and they knew what to expect of a proper hunting community. A few of them lived to tell their story to Knud Rasmussen in 1903 (Rasmussen 1908).

Not only did the small contingent of Baffin Landers contribute vitally to a boosting of the population, they also taught people to use the (now) available wood for bows and arrows, fishing spears, and not least kayaks, about which old Merqussak, one of the immigrants, said, 'we taught them to build kayaks, and to hunt from kayaks. Before that they had only hunted on the ice, and had been obliged during the spring to catch as many seals, walruses, and narwhals as they would want for the summer, when the ice had gone' (Rasmussen 1908: 32). On his side, Merqussak noted how he had been impressed by the quality of the sledges – not least their size and high upright handle posts, which were much better than their own smaller ones.

Through the high-resolution picture of misery rendered by the early explorers in the nineteenth century, we realise both the dire impact of the lack of vital materials during the Little Ice Age and of the new epidemics that the sailors brought with them. In the words of Le Roy Ladurie (1978), the globe had become unified by disease in the fourteenth to seventeenth centuries; germs had spread before (and after) this period, but their circulation intensified with new patterns of trade and with the conquest of the Americas. This unification, or 'the creation – first in Eurasia, and subsequently in the Atlantic area – of a "common market" of microbes, passed through a particularly intense, rapid, dramatic, one might even say apocalyptic phase, during the period roughly 1300–1600' (Ladurie 1978: 30). Until the nineteenth century, the seclusion of the Northwest Greenlanders had protected them from this, but with the new climate they became exposed to new germs. The opening of the sea route brought new resources to a depleted landscape, but it came with a high price of epidemics, and demographic near-extinction. In other words, the ecological crisis might have abated, but soon another dramatic global process infested their life.

Around 1900 they became an object of ethnographic interest, mainly through the reports of Robert Peary (1898) and Knud Rasmussen (1905, 1908). Thus, they entered into an anthropological knowledge economy that allowed Marcel Mauss to sum up their plight in 1906, suggesting that the Eskimos at Smith Sound were in a

> ... miserable state. The expansion of inland ice and the persistence of drifting ice throughout most of the year not only put an end to the arrival of driftwood but obstructed large whales, and made it impossible to hunt whales, walruses and seals in open waters. The bow, the kayak, the *umiak* and most of the sleds disappeared because of a lack of wood. These unfortunate Eskimo were reduced to such circumstances that they retained merely the memory of their former technology. (Mauss 1906 [1979]: 42–3)

Both by insiders and outsiders, there was a strong sense of doom related to the Thule people; they were seen as living on the brink of extinction. This doom was equally related to local factors and global connections, being yet another instance of how all corners of the globe gradually became connected through 'guns, germs, and steel' (Diamond 1997). The gradual warming after the Little Ice Age relieved part of the ecological crisis, but laid people bare to a new kind of apocalyptic experience.

## BREAKING THE HEART: THE COLD WAR COMES TO THULE

Something new happened with Peary's recurrent sojourns in the region (1891–1909); epidemics still came and went, but the recurrent famines abated, due to the regular influx of new materials, including guns, that made hunting easier. In 1898, Peary made a complete census and named 250 inhabitants (Peary 1898, I: 511–14). When he withdrew after having reached his ultimate goal, the North Pole, the Thule Trading Station was established in 1910 on the initiative of Ludvig Mylius-Erichsen and Knud Rasmussen. They had spent a year and a half in the region in 1903–04, with the Danish Literary Expedition to Greenland designed to collect tales and traditions of the Polar Eskimo, as the 'new people' they met were named (Rasmussen 1905).

The Thule Station was to continue and regularise the exchange that Peary had instigated. With a permanent trade station, an annual provision ship, a mission, a school and, eventually, a small hospital, the new people were embraced by the Old World. Compared to earlier times, people were thriving; in exchange mainly for fox fur, they had regular access to wood, rifles, ammunition and utensils of many kinds, to flour, coffee and sugar, to cloth, pots and pans, and eventually to free vaccination schemes. A sense of calm settled, the population stabilised, even if there was still an issue of recurrent epidemics and not least of occupational health hazards (Gilberg 1976). Some of my friends in the area today were children in the 1930s and 1940s and they look back on it as a 'good time', if not without individual struggles, of course. This sense of calm was not to last long, however.

The Thule Station was located on a small headland beneath a vast moraine plain; the headland had been inhabited through the ages, and there was still a small settlement, Uummannaq, referring to the 'heart-shaped' mountain on the tip of the headland. On the south side there was something like a natural harbour, where the Station was built. Attracting people from all over the region, it soon became its capital, the centre where everybody would pass by and live more or less temporarily, when it fitted into their nomadic hunting practices (Hastrup 2009). It was well placed in the middle of the region with access to rich hunting grounds.

Yet, in 1953, the settlement was forcefully relocated from its vital centre – at the heart of their lands. The reason was the establishment of the Thule Airbase, one of many tokens of the Cold War. The case is troubling, not least because of the political smokescreen that concealed the real reason for its creation, even in Denmark. The construction began surreptitiously – at least in relation to the international political community. Locally, it was not easy to overlook. We also have a geographer eye witness in Jean Malaurie, who spent the year 1950-51 in the region, and who came to Thule from a sojourn further north, where he had spent most of his time. He did not believe his eyes, when he looked down on the plain from a mountain that he and his friends had just crossed on sledge. He saw a spectacle that had to be a mirage and continues:

> A city of hangars and tents, of sheet metal and aluminium rose dazzling in the sun among smoke and dust on a plain that yesterday had been empty. The most fantastic of legends took form beneath our eyes.
> 'Takou. Look.'
> Like eruptive pustules risen from the depths of the earth the shapes of containers were lined up along the mountain. Their organ colour made the vision seem even more absurd. We descended the snow slope, which shone before us. Our astonishment became stupor. As far as we could see there were lines of lorries, lifting gear, and mountains of cases. Steel framework raised its great arms to the sky. Along the slopes excavators with enormous jaws scraped away in smoke and steam, removing earth and rubbish, which were vomited into the sea

by continuously moving buckets. The noisy breathing of this town reached us. It was a dull rumbling of ceaselessly turning motors. In the greyness dozens of aeroplanes were circling. One of them, nearer to us, came and went like a great bumble-bee, mixing its personal solemn note with the mounting hubbub. This irruption of civilization, seen from the glacier, was sinister. Our dogs howled like death. Two of them threw themselves on one another. We intervened, but without energy. This return to men was a failure. (Malaurie 1956: 255–6)

The invocation of dystopia still resonates in the region, where the crushing of the core of the hunting grounds has not yet been repaired, even if it has receded into the background. The airbase is still there, while the original inhabitants are not. Most of the turf houses at Uummannaq, where people had lived, ended up being destroyed to prevent people from using them on their way between the northern and southern regions of their land, stopping over on a hunting trip. Even though new houses were eventually built for them in other places, and the central functions of the Thule Station were recuperated in Qaanaaq, the heart was broken. While in some ways just (!) a memory, clearly their old landscape is still visibly ruined.

The Cold War was nothing if not an instance of global, geopolitical overheating, implying a profound militarisation of the Arctic (in the West); one of the tokens was the establishment of the Distant Early Warning (DEW) Line across the American continent and continuing at Thule. This not only implied extensive militarisation of the Arctic at the time, but also morphed into a grand idea of modernisation in the High North that is still part of the geography, even if it partly failed (Farish and Lackenbauer 2009). The DEW Line radar stations 'soon filled local waste bins with chemicals and manufactured items, and, no longer useful, [they] were reclassified during the 1960s as toxic sites' (Doel et al. 2014: 10). In Greenland, Thule was part of this line, but it was soon to develop into something else.

From the outset, the airbase was also designed for scientific, geophysical and meteorological experiments; the former mainly concerned studies of the ice cap, whose depth and plasticity were unknown, and the latter implied experiments to prevent white-outs that often made air traffic difficult if not impossible. The next step was to use some of the results for the construction of an under-ice city that would provide a bulwark against Soviet interests in the region. It was argued that the observed retreat of glaciers and the 'polar warming', which had been predicted by scientists in the 1930s, were a threat to US national security interests (Martin-Nielsen 2013: 52). It was therefore essential to build up a solid, long-range defence to protect the Free World.

> The US was acutely aware that the Soviet's capacity for cold environment warfare had caught the Wehrmacht by surprise and did not want to make the same mistake. 'We must make maximum use without delay of the immensity of the Arctic space to push outposts northward as far forward as possible and capitalize on the resulting manifold military rewards', concluded the Arctic Institute of North America. (Martin-Nielsen 2013: 50)

This was the implicit guiding line for the development of the base into a much more ambitious project of building a city under the ice that was to be powered by a nuclear reactor, and designed to withstand nuclear war (Petersen 2008; Nielsen et al. 2014). Camp Century was dug into the ice cap, some hundred kilometres from the Thule Airbase on the plain, and was a fantastic (and phantasmagorical) engineering project, widely seen as proof of the American power at conquering the Arctic environment – an epic construction of an ideal Arctic laboratory designed to signal invincibility (Nielsen et al. 2014: 455–6). Construction began in 1959; tunnels were made, and the under-ice village – complete with living quarters, church, library and so on – was praised in public as a major feat of American technology. The reactor came last, in 1960, but it was deactivated in 1963 and removed the following year; the constant, slow movement of the ice from the ice cap towards the sea took the engineers by surprise.

> This constant movement meant that tunnels and trenches would narrow as their walls would deform and bulge and that settling might cause tunnel ceilings to give in. Thus in the summer of 1962 the ceiling of the reactor room had drooped so low that it had to be lifted five feet to avoid fatal contact with the reactor. Subsequently Camp Century was reduced to a summer camp in 1964 and abandoned altogether in 1966. The last scientific feat at the campsite was the successful drilling of an ice probe through the Ice Cap to the bedrock 1,390 metres below. (Petersen 2008: 79)

This first-ever ice core retrieved from the depth of ice proved to be an archive of climate data that was to re-scale global climate history and to become a forerunner for many subsequent ice-core projects; this was the most positive outcome. The rest of the experiment was abandoned, and by 1968, the population at the Thule Airbase barely exceeded 3,000 men, having peaked at 10,000 in 1962, and the base was transformed from an offensive to a defensive space (Martin-Nielsen 2013: 65). Meanwhile, the people who had been ousted from their hunting grounds still numbered fewer than 400, relying on the landscape in its entirety. On top of the relocation, the nuclear dream finally destroyed the integrity of their hunting grounds.

The stories multiply from there and remind us how deeply the heated geopolitical situation had impinged on the global mind, barely noticing what went on. Some years back I had a long conversation with a seasoned hunter, who had a past as a local political representative in the national government; while out camping by the ice edge for hunting, quite naturally we talked about the changing ice conditions. He said that it was all down to the Americans, who had tampered with the ice cap, making tunnels in it, believing they could make an under-ice road all the way to the south (!). 'They were mad, and they destroyed the natural ways of the ice; look how much water runs out under the glaciers and make them break off so dramatically.' While the causal relation is dubious, the hunter is actually pointing directly to the ruination of their land.

Other inhabitants voiced a fear of there being nuclear waste still around in the now inaccessible Camp Century corridors that may eventually break off from a glacier now nearing the sea; nothing can withstand the force of this. The fear of radiation was reinvigorated by the crash of a B52 plane in 1968, carrying plutonium bombs, disappearing through the sea ice, and creating a minor (allegedly localised and temporary) catastrophe of contamination in the fjord by the old settlement – adding to the toxic waste from the DEW-radar station. Three bombs were retrieved, but a fourth bomb is still in the fjord, it is rumoured locally, if vehemently denied by the authorities, and the latent fear of radiation refuses to go away in spite of negative clinical tests (Bjerregaard and Dahl-Petersen 2010). The ecological crisis lingers; the ruined landscape cannot be obliterated.

## BECOMING HERITAGE: CLIMATE CHANGE AND RE-OBJECTIFICATION

Today, a new and no less poisonous ecological crisis is manifest in the industrial contamination of the sea – a by-product of new extractive industries in the Arctic, and of global pollution in general. This afflicts the entire food chain, concentrating in the marine mammals, the top-predators hunted and eaten by people, in spite of their being duly warned off eating too much of it (see, for example, AMAP 2002, 2011, 2015). As the Thule community has embraced the new world of communication technologies over the past ten years or so, when satellite senders have broken their isolation in a new way, they are acutely aware of this new global threat to their most cherished food.

This knowledge accompanies the high-profile discussion of climate change, experienced directly with the sea ice being ever more unpredictable, thus compromising access to hunting grounds that have served people well for a long time. The game also seems to seek out new territories, sometimes beyond reach, and there is a definite sense that the liveable space is shrinking. Ancient sledge routes that connected settlements to each other, and to the distant haunts of walrus, polar bear or narwhal, the emblematic species, are disappearing. The all-important seals have served as a stable crop throughout the year, either hunted in open waters during summer, by their breathing holes on the sea ice during spring, or by nets lowered though holes in the sea ice close to the settlement during the period of winter darkness; the latter is now compromised by the unstable ice, claiming both seals and nets if the ice breaks up. Seabirds, muskox and, occasionally, reindeer may likewise make a difference in the natural economy, but they are not as essential and often equally inaccessible. Fish – polar cod, arctic char and halibut – also add to the diet, the latter increasingly so, because it has become an object of trade – and one that does not necessarily demand long-distance travelling.

Living conditions are strained, meaning that some people may go hungry during winter; they cannot afford to buy supplies from the store. The diminished hunt also affects the dogs, who need a lot of real meat – bear, walrus or seal – to have the strength to drag the heavy sledges. If not properly fed over winter, the dogs may succumb, and leave the family with even less capacity to hunt in the following spring. All while the environment is changing, the materials and technologies by which people have lived for ages may gradually become obsolete, including their spectacular skin clothes. Small wonder that international film crews, journalists and researchers flock to the main town of Qaanaaq to document 'The Last Eskimos' – as popular film titles often have it, be they in English, German, Japanese or Korean. The inhabitants of the region are already figuring as lost and turned into heritage for protection.

'Heritage is our legacy from the past, what we live with today, and what we pass on to future generations. Our cultural and natural heritage are both irreplaceable sources of life and inspiration' – as UNESCO has it on World Heritage (1972). It comprises sites (including archaeological sites) that are of outstanding universal value from the historical, aesthetic, ethnological or anthropological point of view (see Article 1 of the UNESCO Convention). Truly, the hunting life is an irreplaceable source of life and inspiration, and the sites of the Thule culture are certainly of outstanding value from all points of view. Yet, if climate, politics and science have already changed the place to a prime example of an Anthropocene world, how can social life be protected against this? And how may being portrayed as a legacy from the past affect the local community's pride and agency in face of the massive challenges? Recently, a process to make the region into a UNESCO Biosphere Reserve has been set in motion, to protect the habitat of important marine species. This may eventually temper the intensifying ocean traffic in the region, and hunters may still have controlled access to important hunting grounds, but the larger issues of industrial pollution and chemical contamination hailing from far away cannot be solved this way – and the animals will become increasingly inedible.

Making particular sites and their human and non-human inhabitants subject to protection implies a kind of objectification that may be misplaced, due to their singling out of the particularities of the site as *things in themselves*, while in fact totally heterogeneous and made up of an endless array of things (Kopytoff 1986: 70). The general point is that neither objects nor properties are independent of a subject. Cultural heritage has no ontology; it is made up of an ever-unfolding life.

For the inhabitants in Northwest Greenland, their life is for real. The sledge is made for travel, not for exhibition – such as happened to the sledge John Ross retrieved from the Arctic Highlanders in 1818 and donated to the British Museum. The formidable dog sledges are still the predominant mode of travel, and the primary instrument of hunting and of community-making – allowing people to move about and visit each other between the settlements. With the dwindling ice, both of these activities have become more circumscribed by the increasingly unreadable ice (Hastrup 2013), and people begin to look for alternative futures. Meanwhile, film crews, journalists and wildlife tourists are pouring into the community, especially in May, when the cold is abating under the midnight sun, and there may still be some hunt from the ice.

This presents the hunters who still have enough dogs for extensive trips with a dilemma; should they go for real, if uncertain, narwhal hunting at the ice edge, or should they collect a payment in cash for taking out tourists, 'performing' as hunters. This is a delicate balance, not least because they often find tourists completely unprepared for the hardships of travel by sledge, and for the fact that one cannot 'book' an animal to show itself at a particular time to fit into the outsider's brief visit. Another dilemma has arisen between going for the new cash crop in the region, the halibut, or going for bigger game, further away – which is more uncertain but also feels more right, if your strength and your dogs are intact. The halibut can still only be caught from the ice on long lines, the same ice that may carry the hunters further afield. While increasingly destabilised, the ice remains the all-important infrastructure, with nothing as yet to replace it; choices must be made. During winter, when darkness reigns (there is no sun for four months) there is less choice and no tourists. In this season, people seem partially to hibernate. A generation ago there would have been communal walrus hunts in early winter, when the new ice was stable enough for people to walk on it, silently on their soft *kamik*-soles, yet also still frail enough for walruses to break through it from below with their massive heads in order to breathe; this is a thing of the past now, because the ice does not stabilise until mid-winter, mostly in January now, when it is too dark for any hunt of that kind. Along with the communal hunting, the sharing of hunting stories during the winter night has likewise become a thing of the past, even if a lucky anthropologist may still be offered a replay on the spur of the moment, as it has been my privilege to experience.

My contention is that outsiders, potentially including anthropologists, rushing to the place to experience 'the last' hunters with their bearskin trousers, sealskin *kamiks*, dog sledges, kayaks and harpoons, contribute to an image of life as if already just an exhibit. Politics evidently enters into the equation here, as the link between value and possibilities for exchange (Appadurai 1986: 57). We are reminded of the shifting values attached to medieval relics, whose apparently miraculous powers were renewed, and often upgraded, as they were moved to another sacred place (Geary 1986). The movement now is virtual and mostly digital, but the circulating relics from a disappearing world still shift the value of its 'things' around. The price is a loss of people's contemporariness, their modern and well-informed outlook, and not least their singular capacity for seizing the moment, any moment of potentiality.

Another price is an objectification, a new essentialism of Inughuit culture, a name neither of absolute local currency, nor old enough to be anything but a statement – which they make on their own behalf from time to time, especially when addressing their politicians. Outsiders, using it uncritically as 'their name', contribute to a cultural essentialism, which cannot be sustained. The inhabitants are a mixed group of people from old families, mixed marriages and temporary employees. They may be connected through place, but their individual histories mock any idea of their thinking alike. Some people want to remain in the region, and others want to move somewhere else; some prefer the hunting life, others look forward to possible mining adventures in the North. Seeing people as living heritage dethrones them from their own time and individuality.

## ARCTIC OVERHEATING: CLOSING THE ARGUMENT

This chapter has shown how 'discovery' and 'doom' were part of the same picture in the nineteenth century, when extinction of the newly-found people was almost constitutional. They were victims of the Little Ice Age and its impact on their mobility, and soon became prey to a common market of discovery and disease that almost exterminated them, having been reduced to a point where they were barely able to reproduce. In the twentieth century, the Cold War in particular scarred their landscape and destroyed the integrity of their hunting grounds. Presently, new forms of overheating, both in the sense of global warming and of ocean contamination, undermine hunting life, which becomes fixed as heritage, all while locally, new imaginative horizons open up (Crapanzano 2004) – including a (vague) dream of possible mining developments. Meanwhile, outsiders' territorial interests in the Arctic intensify as new possibilities for extractive resources emerge; whatever the answers to these claims, they are crucial both to geo-strategic planning and to national mythologies in the involved states (Doel et al. 2014: 4). In an odd way, and despite the natural economy that has governed life in Thule, the dire state of the deep Arctic seas and the territorial struggles over the Polar region show that even in the quintessential wildness, life unfolds in capitalist ruins (Tsing 2015). Science has not been innocent in the process of ruination.

What we are faced with in the Thule Region is a reconfigured 'imperial' landscape, pointing towards Ann Laura Stoler's question: 'How do imperial formations persist in their material debris, in ruined landscapes and through the social ruination of people's lives?' (Stoler 2008: 194). The answer is not to fall back on colonial criticism (this is implicit), but to show the long-term, elusive effects of modern processes of ruination and wasting through proper analysis. This is where the dire effects of overheating transpire.

## ACKNOWLEDGEMENTS

Recurrent fieldwork in the Thule Region was made possible through the project 'Waterworlds' (2009–14) funded by the European Research Council (Adv. Grant 229459). Later, the interdisciplinary 'North-Water project' (2014–17) allowed me to continue in the field and to embrace new perspectives. Thanks to the Carlsberg Foundation and the Velux Foundations for generous financial support.

Thanks also to Thomas Hylland Eriksen and Astrid Bredholt Stensrud from the 'Overheating' project for organising the conference at which a first draft of this chapter was presented, and for their vital comments on later versions.

REFERENCES

AMAP 2002. *AMAP Assessment 2002: Persistent Organic Pollutants in the Arctic.* Arctic Monitoring and Assessment Programme (AMAP), Oslo, Norway.
—— 2011. *AMAP Assessment 2011: Mercury in the Arctic.* Arctic Monitoring and Assessment Programme (AMAP), Oslo, Norway.
—— 2015. *AMAP Assessment 2015: Human Health in the Arctic.* Arctic Monitoring and Assessment Programme (AMAP), Oslo, Norway.
Appadurai, A. 1986. Introduction: Commodities and the Politics of Value. In A. Appadurai (ed.), *The Social Life of Things. Commodities in Cultural Perspective.* Cambridge: Cambridge University Press, pp. 3–63.
Bjerregaard, P. and I.K. Dahl-Petersen. 2010. *Sundhedsundersøgelsen i Avanersuaq 2010.* København og Odense: Statens Institut for Folkesundhed.
Bravo, M.T. 1998. The Anti-anthropology of Highlanders and Islanders. *Studies in the Historical and Philosophical Sciences* 29(3): 369–89.
Crapanzano, V. 2004. *Imaginative Horizons: An Essay in Literary-Philosophical Anthropology.* Chicago, IL: University of Chicago Press.
Diamond, J. 1997. *Guns, Germs, and Steel. The Fates of Human Societies.* New York and London: W.W. Norton & Company.
Doel, R.E., U. Wråkberg and S. Zeller. 2014. Science, Environment, and the New Arctic. *Journal of Historical Geography* 44: 2–14.
Farish, M. and P.W. Lackenbauer. 2009. High Modernism in the Arctic: Planning Frobisher Bay and Inuvik. *Journal of Historical Geography* 35: 517–44.
Geary, P. 1986. Sacred Commodities: The Circulation of Medieval Relics. In A. Appadurai (ed.), *The Social Life of Things. Commodities in Cultural Perspective.* Cambridge: Cambridge University Press, pp. 169–91.
Gilberg, R. 1976. *The Polar Eskimo Population, Thule District, North Greenland.* Copenhagen: Nyt Nordisk Forlag Arnold Busck (Meddelelser om Grønland, 203/3).
Hastrup, K. 2007. Ultima Thule. Anthropology and the Call of the Unknown. *Journal of the Royal Anthropological Institute* 13: 789–804.
—— 2009. The Nomadic Landscape. People in a Changing Arctic Environment. *Danish Journal of Geography* 109(2): 181–9.
—— 2013. Anticipation on Thin Ice: Diagrammatic Reasoning Among Arctic Hunters. In K. Hastrup and M. Skrydstrup (eds), *Anticipating Nature. The Social Life of Climate Models.* London and New York: Routledge, pp. 77–99.
—— 2017. The Viability of a High Arctic Hunting Community: A Historical Perspective. In M. Brightman and J. Lewis (eds), *The Anthropology of Sustainability. Beyond Development and Progress.* New York: Palgrave Macmillan, pp. 145–64.

—— 2018. The Historicity of Health: Environmental Hazards and Epidemics in Northwest Greenland. *Cross-Cultural Research*. https://doi.org/10.1177/1069397118806823.

Hayes, I.I. 1866. *The Open Polar Sea. A Narrative of a Voyage of Discovery Towards the North Pole, in the Schooner United States*. London: Sampson Low, Son and Marston.

Kane, E.K. 1856. *Arctic Explorations. The Second Grinnell Expedition in Search of Sir John Franklin, 1853–55*, vols I and II. Philadelphia: Childs & Peterson.

Kopytoff, I. 1986. The Cultural Biography of Things: Commoditization as Process. In A. Appadurai (ed.), *The Social Life of Things. Commodities in Cultural Perspective*. Cambridge: Cambridge University Press, pp. 64–91.

Ladurie, E.L.R. 1978. A Concept: The Unification of the Globe by Disease (Fourteenth to Seventeenth Centuries). In *The Mind and Method of the Historian*, trans. S. and B. Reynolds). Brighton: The Harvester Press, pp. 28–83.

Malaurie, J. 1956. *The Last Kings of Thule*. London: George Allen and Unwin.

Martin-Nielsen, J. 2013. 'The Deepest and Most Rewarding Hole Ever Drilled': Ice Cores and the Cold War in Greenland. *Annals of Science* 70(1): 47–70.

Mauss, M. 1979 (with H. Beuchat). *Seasonal Variations of the Eskimo: A Study in Social Morphology*, trans. and introduction J.J. Fox. London: Routledge and Kegan Paul. Originally published 1906.

Nielsen, K.H., H. Nielsen and J. Martin-Nielsen. 2014. City Under the Ice: The Closed World of Camp Century in Cold War Culture. *Science as Culture* 23(4): 443–64.

Peary, R.E. 1898. *Northward Over the 'Great Ice'. A Narrative of Life and Work Along the Shores and Upon the Interior Ice-cap of Northern Greenland in the Years 1886 and 1891–1897*, vols I and II. London: Methuen & Co.

Petersen, N. 2008. The Iceman that Never Came. *Scandinavian Journal of History* 33(1): 75–98.

Pratt, M.L. 1992. *Imperial Eyes. Travel Writing and Transculturation*. London: Routledge.

Rasmussen, K. 1905. *Nye Mennesker*. Copenhagen: Gyldendal.

—— 1908. *The People of the Polar North*. Compiled from the Danish originals and ed. G. Herring. London: Kegan Paul, Trench, Trübner & Co.

Ross, J. 1819. *Voyage of Discovery, Made Under the Orders of Admiralty, in His Majesty's Ships Isabelle and Alexander, for the Purpose of Exploring Baffin's Bay, and Inquiring into the Probability of a North-West Passage*. London: John Murray, Albemarle Street.

Stoler, A.L. 2008. Imperial Debris: Reflections on Ruins and Runination. *Cultural Anthropology* 23(2): 191–219.

Tsing, A.L. 2015. *The Mushroom at the End of the World. On the Possibility of Life in Capitalist Ruins*. Princeton, NJ and Oxford: Princeton University Press.

UNESCO. 1972. *Convention Concerning the Protection of the World Cultural and Natural Heritage*. Paris.

# 4. Sea Ice, Climate and Resources
## The Changing Nature of Hunting Along Greenland's Northwest Coast

*Mark Nuttall*

For the hunters and fishers who live in the Upernavik district and Melville Bay area of Northwest Greenland, the activities of daily life are finely tuned to a world in motion; to the vagaries and extremes of the weather, to seasonal variations and the transition from winter darkness to summer light, to the migrations and local availability of marine mammals and fish, to the power of the sea, tides, currents and wind, to the formation and permutations of sea ice, to the cracking, rattling and rumbling of tidewater glaciers, and the drift of icebergs that calve from them. But the Arctic is heating up and people are confronted with a range of new challenges arising from quite dramatic changes in weather and climate (including considerable fluctuations in precipitation patterns and snow cover), to the sea ice and coastal waters, and to the spatial distribution and behaviour of animals. Many of these challenges are practical and economic, as well as affectual and political, and they have implications for how people move around, engage and interact with their surroundings, and how they make a living.

The Arctic has never been an inert, empty space, despite what many popular stereotypes of it as a timeless, vast frozen wilderness otherwise suggest. A region of flux, impermanence and becomings, it has experienced tremendous shifts in climate over millennia. Along with contemporary indigenous and local observations of environmental transformation, a considerable body of recent scientific work shows how circumpolar ecosystems are becoming far more turbulent than has been known in the recent history of the Arctic and, possibly, approaching tipping points (for example, Lind et al. 2018). Circumpolar seas in particular have become environments of scientific concern as the uptake of anthropogenic $CO_2$ increases the acidification of the Arctic Ocean (Di et al. 2017) and as global warming continues to push the region towards what many scientists worry will be an ice-free future (Screen and Williamson 2017; Wilkinson and Stroeve 2018). At the same time, many parts of the Arctic are attracting increased interest as sites for extractive industries, cruise ship tourism is expanding into areas previously thought of as remote and difficult to navigate through, especially in places where ice has usually impeded maritime transit, and conservation organisations and environmentalist groups place iconic species such as polar bears and whales up front and centre in global campaigns to save and protect circumpolar lands, waters and ice.

As the effects of climate change and resource development, indigenous rights, conservation, sovereignty, and environmental and political security in the global North attract greater attention, Greenland in particular has moved into sharper international focus over the last decade. This self-governing territory of the Danish Realm is often at the forefront of scientific and media reports about a melting Arctic and the Anthropocene. It has also come under the speculative gaze of a larger number of mining companies, oil and gas explorers, and their investors, who are encouraged in their ambitions to probe the subsurface by Greenland government institutions responsible for developing an extractive industry sector. Ice may be disappearing, and coastal waters may be warming, but large areas of land and great stretches of seabed are being defined and marked off as spaces of resource abundance with significant economic potential for global capitalist enterprise. The physical (and the imagined) properties and materialities of minerals and hydrocarbons, and the places in which they are located, have become key to anticipatory resource economics, but they also enter into energetic and often fraught political and public discussion about Greenland's future and the territory's relationship to Denmark, as well as resource governance, regulatory procedures and environmental management (Nuttall 2017). Smaller communities are also caught up in these discussions, as economic development plans, including those for extractive industries, new airports, harbours and industrial zones are mapped out in municipal and national strategies. The challenges arising from environmental transformation combine with political and economic ones, as well as broader processes of social change. In this chapter, and providing examples from long-term anthropological fieldwork in the Upernavik and Melville Bay area, I consider some aspects of how all of this has a bearing on the region and on people's everyday lives.

## NORTHWEST GREENLAND IN AN OVERHEATING WORLD

Arctic temperatures are expected to increase several times that of the global mean in the coming decades (IPCC 2013, 2014). Already, though, in an overheating world the shifting geographies of high latitude regions are increasingly apparent through diminishing glacial ice, the retreat and thinning of sea ice during winter as well as summer, thawing permafrost, coastal erosion, an intensification of stormy weather, and changes to the migration routes and population sizes of a number of animal and fish species, such as whales, polar bears, caribou, cod and mackerel (ACIA 2005; AMAP 2012). The seasonal melting processes affecting the edges and the surface of parts of Greenland's immense inland ice are increasing in summer (Box and Decker 2011), while summer sea ice is decreasing in the Arctic Ocean and other parts of the High Arctic at a rate faster than foreseen in any climate model. Many scientists have been warning for some years that this perennial summer ice may fully disappear before 2050 (for example, Wang and Overland 2009; Stroeve et al. 2012; Wadhams 2012). But observations of decreasing winter ice cover of the Arctic Ocean now alarm scientists as much as the lower summer extent – in March 2018, for example, scientists from the NASA-funded National Snow and Ice Data Center (NSIDC) in Boulder, Colorado, recorded the second lowest maximum extent for winter seasonal ice since satellite records began in 1979.[1] Their assessment was that the ice formation and growth season had ended with extremely low sea ice extent in both the Pacific (the Bering Sea) and Atlantic (the Barents Sea) sides of the Arctic, and that this reduced ice cover reflected a combination of influences, including a late autumn season freeze-up and high air temperatures that persisted throughout the winter. The frozen Northern seas of today, scientists and environmentalists point out with increasing regularity, and with reference to such data, are likely to be the open (or at least, less ice-filled) waters and new sea lanes of the not-too-distant future. This appears to be borne out by the more commonplace transit of cargo vessels and liquefied natural gas tankers through Russia's Northern Sea Route, some of them making the journey from Europe to eastern Asia, or vice versa, navigating Eurasian Arctic waters and, increasingly, without assistance from icebreakers.

In the northwestern part of Greenland, along a sizeable stretch of Baffin Bay coast comprised of numerous islands, headlands, fjords and glaciers, some 1,100 people live in the town of Upernavik with a further 1,700 or so living in ten small communities (including Savissivik in Melville Bay). Most households in this region depend on hunting and fishing for their livelihoods, with social, cultural and economic life organised around strong and enduring patterns of social relatedness and kinship which entwine and embed human interests with the other than human in dynamic surroundings of ice, water and land (Nuttall 2009). While hunting marine mammals, such as seals, belugas and narwhals, and catching several species of fish are activities that procure food for the household and community, as well as for distribution and sale through local and national networks, a commercial, albeit small-scale, fishery for Greenland halibut has developed since the late 1980s. This provides much of the income for many people in the district, but is often considered by them to be a way of supporting and sustaining hunting rather than replacing it completely.

This cryospheric world is in perpetual motion and the Arctic has always demanded of indigenous people an acute awareness of the importance of being alert and prepared for the unexpected (for example, Briggs 1991; Nuttall 1992). As hunters always remind me when I travel with them on the sea ice or open water, anything can happen unexpectedly and come down on you (*terlassivaa*), whether it is a storm, a sudden swell or a walrus rising from the depths of the sea and capsizing your boat. Feeling safe – *terlik* – is only a reminder of insecurity and *terlarorpoq* means that if you do something, or go out on your dog sledge or set off in your boat, thinking you are safe, then you are unaware of how your actions can actually bring danger and misfortune. Greenland's extensive coastal areas have been inhabited by various human populations for some 4,500 years and the country has experienced several episodes of human migration, settlement, colonisation, resource exploitation, cultural transitions and environmental change during this long period. In Northwest Greenland access to places to hunt marine mammals from sea ice (and the ice, or floe edge) or during the ice-free water summer season, as well as swift and efficient travel over frozen surfaces as much as choppy seas, has been a prerequisite for the survival of historic and contemporary communities. Assessing and determining just how far climate change has had an influence and impact on past transitions and adaptations, though, is hampered by a lack of quantitative and continuous paleo-climatic data that can be beneficial for an understanding of historic climate and weather events near sites of former seasonal and more permanent settlements and animal migration routes. However, long-term engagement with the Arctic, and with the non-human entities and materials that comprise and fill it, and which are constitutive of it, such as marine mammals, fish, rocks, water, coastlines, ice, snow, wind and so on, has required people to have a high degree of mobility as well as flexibility and versatility in social organisation, settlement patterns and resource use practices (Nuttall 1992; Petersen 2003; Hansen 2008). To live along the coasts of this Northern world has required effective and creative responses to environmental, economic and social change at many times in the past, but also an ability to anticipate and map out the region's potential, exploring places to hunt, fish, live and raise families.

In Northwest Greenland, sea ice (*siku* in kalaallisut (Greenlandic)) is a dominant feature of people's lives for several months of the year. Its gradual formation in winter as the surface of the sea freezes, congeals and solidifies brings all kinds of possibilities for hunting, fishing and travelling. It allows for the creation of an intricate and extensive pattern of trails, enabling travel by dog sledge on the sea ice, but also from it, across snow-covered land, and up on to glaciers. This widens access to places to hunt and connects communities by very different routes than the open water does. At the same time, the apparent solidity and enduring presence of *siku* through a long winter and into spring is deceptive. I have never lost my sense of wonder about being able to walk on and travel across ice, but this is countered by an ever-present understanding that one can slip through it unnoticed and disappear into the freezing cold, dark water below. Sea ice can be many things. Firm, dense, tensile, soft and malleable, it can separate into a shifting pack without warning, tides and winds can pull large sections of it apart, its surface can liquefy and turn to slush, and headland cracks can suddenly open up. People are aware that the fixity and fastness of *siku* cannot and should not be taken for granted. *Siku* is hard yet brittle, just as soapstone is (*sikappoq*). When the surface of the sea freezes, one has always to remember that it has merely been glazed (*sikassarpaa*) by *siku*. Hunters talk of the need to *qajassuuppaa* the ice – which means to approach it, travel on it, and camp on it cautiously, and to be careful with it at all times.

The need for this sense of caution – and awareness of the fickle nature of ice and its various states and qualities – seems to be talked about in the communities of Northwest Greenland as being far more essential in a time of rapid climate change (Nuttall 2017, 2018). People have been witness to how sea ice has been undergoing a trend of thinning, and how its extent has been dwindling each winter and spring, for the last couple of decades. In February 2017, for example, friends in the village of Kangersuatsiaq, which lies to the south of Upernavik, told me they were anxious about large holes in the ice; they described them as wounds that were slow to heal. Thinner ice, or worse, no ice at all, makes travel difficult and reduces the options people have to mark out sledge routes to hunting places and between communities. Glaciers too are receding with dramatic effect, reshaping local topographies. Fjords and bays are choked with the icy rubble from collapsing glacial fronts, icebergs appear to be bigger, block channels, bays, harbours and islands, and extensive tracts of land are uncovered as the glaciers move back and leave the debris that shape lateral and terminal moraines. Increased meltwater runoff from those same glacial fronts is affecting water temperature and circulation patterns as well as the formation and durability of sea ice. This influences the distribution, presence and availability of marine mammals and fish that provide the basis for local livelihoods and household economies. The absence of ice, or, when it is present, its increasingly weakened structure and its softer and far slushier texture, makes travel by dog sledge or snowmobile almost impossible in many parts of the district, restricting the distances people are able to move and travel around during winter and spring. Alternative dog-sledging routes over glaciers are being explored in some parts of the region, but even then, access to them is becoming difficult because of both the shifting sea ice regime and changes to glacial ice.

In the town of Upernavik, one of the most striking differences I notice on visits there compared to say 15 or so years ago is that, apart from the observable changes to sea ice cover and extent in winter and spring, the rhythmic flow of daily life plays out against the silence of dogs. More accurately, though, by this I mean it is the absence of Greenlandic sledge dogs (*kalaallit qimmiat*) that one feels and notices more than perhaps any other change. They have become fewer since the early 2000s. The air may feel moister, it may rain more in winter, spring and summer, new houses have been built, and people have been relocating to the town from the settlements in the district, but one friend who had moved some years ago to Upernavik from Kangersuatsiaq told me that, above all, she felt that the atmosphere of the town has an indescribably sad feeling now that most conversations between *inuit* (people) and their *qimmiat* have been silenced. Like the smaller villages in the district, for the hunters and fishers of Upernavik dogs have been vital for transportation on sea ice. *Qimmeq* is the singular for sledge dog and means 'one that pulls' (*qimuppoq* means to pull the sled); and while *qimmit* is the plural form, *kalaallit qimmiat* is the specific name for Greenlandic sledge dogs. But a hunter's *qimmiat* do more than pull sledges – they form and nurture skilled hunting partnerships with their human companions. These partnerships are essential for success in hunting – for example, a polar bear hunter (*nannunniaq*) cannot catch a bear on the sea ice without his dogs who at a crucial moment during the pursuit are released from the *pitoq*, the strap on the front of the sledge to which their traces are attached, and become *nallertut*, those that surround the bear. As an aside, *qimusseq* is the word for a sledge with a person or people on it – something that is being pulled by dogs, the activity of dog sledging – while *qamutit* is the assemblage of a sledge with dogs and humans. Yet, as in other parts of the district, for a number of years now the ice has not been forming as solidly in the marine environment surrounding the town (which is located on an island) in ways that make for safe travel and access to the places people need to reach to hunt and fish. When I first travelled to Upernavik in the late 1980s, some 1,500 sledge dogs made their presence known by a constant howling. Most hunters have now given their animals away to relatives and friends in the settlements, or have put them down. Only a handful of hunters keep a full dog team, while others keep one or two dogs for transporting small sledges around town when moving supplies or transporting fuel, or when they can get out on the ice to do some fishing close to the community. The construction of new houses with little space between neighbouring lots (together with municipal regulations for where and how one can keep dogs) has also meant there is no room for some hunters to chain dogs outside their homes should they wish eventually to rebuild a team. Along with this decline in keeping dogs comes a loss of skills, in how to handle, engage, work and collaborate with their *qimmiat*, how to nurture the leader of a team (*qimmit ittuat*), how to consult with the oldest dog in the team (the *ittoqut*), and also a loss of the finely grained knowledge and detailed linguistic terminology necessary for building a sledge and making other equipment, such as harnesses and whips. As one friend who had once kept a team of 16 dogs explained to me, to give them up, to lose them, was as if they had gone adrift with the disappearing ice, and been carried far out to sea. And for those who do still keep a full team, the costs incurred in ensuring a steady supply of food (fish, seal or walrus meat, or dry food bought from the local store) for their dogs place a burden on household expenses at a time when opportunities for engaging in hunting and fishing activities have been reduced.

The changing nature of the sea ice around Upernavik, as well as the decline in the number of dogs, has had an impact on the livelihoods of people who have moved to the town from the villages. In May 2017, two former residents of Kangersuatsiaq – people I have known since the late 1980s – spoke to me about how they had moved to Upernavik because they thought prospects for making a living by fishing would be better than if they had remained in the village. However, while other migrants from Kangersuatsiaq have managed some success with fishing or have found employment, these two men have given up hunting and fishing in the last few years. They ascribe this partly to poor sea ice conditions and an inability to be able to get on either the ice or water because of the difficulty of leaving the harbour. They now both work making carvings for sale to tourists who visit on the few cruise ships that make a brief stop in Upernavik during summer. The nature of this work is seasonal and brings in little income, and so they seek to supplement this with casual, part-time employment with the municipality.

### THE CHANGING NATURE OF HUNTING

In the Upernavik district, hunting and fishing are not just affected by melting ice or stormier weather. When I first began fieldwork in

Northwest Greenland in the late 1980s, people's livelihoods and hunting activities had been disrupted severely by anti-seal hunting campaigns during the decade as well as by the EU (then EC) Sealskin Directive, which had imposed a ban on the import of many products derived from the seal hunt into Europe. During my more recent research in the area, I have witnessed how the collapse in global markets for sealskins has left an unwelcome legacy for many households over 30 years later (even if that ban has been relaxed somewhat). International opposition to hunting marine mammals continues, but in less immediately confrontational ways, and public distaste for furs ensures that sealskins are not a material of choice for coats, hats, gloves and purses. Stronger conservation measures, management regimes and quota systems set in place by Greenland government departments as well as international agencies in response to scientific concern over declining whale, walrus or narwhal and beluga populations, as well as worries about overfishing, also act to marginalise hunting and commoditise animals and fish as tightly regulated resources. Quota systems, for instance, place limitations and restrictions on hunting and fishing activities, with implications for social life and patterns of sharing meat and fish, but they also contribute to changes in attitudes towards animals and the environment. A transition from seal hunting to halibut fishing began in the late 1980s and early 1990s, partly as a response to the loss of income from selling sealskins but also as a way of providing a form of economic development for the villages in Upernavik district. Many people in the district would agree that this has brought economic advantages and enhanced the possibilities to earn money, to build new houses and improve living conditions.

Fishing has, though, also reshaped demographic patterns to a considerable extent as people continue to move to the town of Upernavik or to other villages where fishing prospects are better and which have the infrastructure and capacity for buying, freezing and storing fish – but, no matter if for some people fishing is carried out as a way of supporting marine mammal hunting, it has also meant the decline of traditional subsistence culture and, to some extent, the fragmentation of social life built around kinship, social relatedness and community (Nuttall 1992, 2017). Fishers must locate themselves near the best fishing grounds, as well as the places with facilities to land their catch, and many often spend several months of the year away from their home communities. Most villages in the Upernavik district have fish landing and freezing facilities managed by Royal Greenland, a Greenland government-owned company focused on fishing and fish processing, but communities such as Naajaat and Kangersuatsiaq do not (the facility in Kangersuatsiaq closed in 2011 because of a lack of fresh water), meaning fishers from there must travel to villages such as Kullorsuaq, Nuussuaq, Nutaarmiut, Innaarsuit and Tasiusaq in the more northerly parts of the district to fish and land their catch. Some even make a more or less permanent move to those communities, settling there because of the better possibilities for fishing and selling Greenland halibut.

Regulations, quotas and management systems for living marine resources are decided upon and implemented by Greenland government departments in the country's capital, Nuuk (often in association with, or adhering to, the rules of broader international regulatory agencies). Management regimes are based on scientific information about the health and status of fish stocks and the population estimates for large marine mammals such as polar bears, narwhals, walrus and other species of whale, which derives from research carried out mainly as part of the assessment and monitoring work of the Greenland Institute of Natural Resources (GINR) in Nuuk – an advisory body to the self-rule authorities. These regulations have nothing to do with climate change explicitly, or because of scientific worries over climate-induced influences on marine and terrestrial wildlife, but are responses to advice about the status of animal populations (which are, nonetheless, increasingly affected by environmental and climatic changes) and the level of hunting or fishing that should be allowed. Local hunters complain, however, that such assessments are carried out by scientists based at research and environmental governance institutions in Nuuk or Denmark with no consideration of local knowledge about marine mammals or fish. The effects of such management systems combine with changes in environment and climate to influence, determine and impact the abilities of people to hunt and fish certain species. Often, people feel it is harder to negotiate and comply with these regulations than adapt to, or anticipate, changing sea ice conditions. In many cases, rather than being the predominant concern for people, climate change intensifies the societal, political, economic, legal, institutional and environmental challenges that already affect everyday life in northern communities. As one hunter and fisher from Naajaat, a small settlement north of the town of Upernavik, put it to me in summer 2015 (and I find his comments on warmer weather and oil exploration, as well as the view that hunters are often considered to overexploit resources, illustrative of local perceptions and experiences of the complexity of interrelated changes):

> As a fisher, I have noticed some years there are fewer halibut and some years there are more. Our theory is they're leaving for further west to the open sea. If the seismic activities cause damage to what the halibut eat, the halibut will also be damaged.

Is the warm weather going to keep getting warmer? Is it going to be like this forever? I read a report by some Greenlandic scientists that the climate fluctuates every 100 years or so. Old hunters talk about how they experienced similar changes, and also saw foreign species in coastal waters, at some points in their lives too.

The one in charge of nature is nature itself. There are already regulations that nature imposes. The environment is already being protected by the way nature works. For example, a few years ago there were a lot of ducks. The sea suddenly froze so the ducks went to Upernavik. They crashed into lights and buildings and many died. There must be a lot of ducks already dying in the sea ice. There was a similar encounter with red fish a few years ago. A lot of red fish had died and were just floating in the sea because of some natural changes. It is very difficult and frustrating to be living with political regulations when the animals are just dying anyway and sometimes we get the blame for it.

Hunters also point to the difficulties they often encounter as a result of the inflexibility of the rules for holding a full-time hunting licence. These require a hunter to earn 50 per cent or more of his income from hunting and fishing activities. While hunters may spend all their time hunting and fishing, their income is relatively low as only some meat and fish are sold; the rest being used by and for the household, as well as moving within local networks of sharing and distribution. As such, hunting as a livelihood is geared mainly towards the household and procuring food for household and community consumption. It is this characteristic which makes hunting a form of subsistence activity that sustains a social economy rather than being a market-oriented occupation. A hunter could earn a reasonable, occasional income to supplement hunting and fishing if tourists or scientists wished to charter his boat and employ him as a guide for a few days or a week, but this money could easily exceed 50 per cent of his total earnings. In such a case, he would lose his licence and would be prevented from hunting those animals which are subject to a quota, such as polar bears, narwhals, belugas, walrus and other whales.

In some ways, the changing nature of hunting is characterised by the contrast between a stubborn persistence of older hunters to cling on to particular ways and traditions, and who try to resist moving away from catching marine mammals to fish for a living, and the demands placed on younger hunters to participate in more commercial forms of harvesting and become fishers. To hunt is to strive for something, to want something, and to go after it (*piniarpaa*), and to be a hunter is to be a provider. But fewer hunters (*piniartut*) today are said by the older generation to be like the good, capable hunters (*pinattorittut*) of the relatively recent past. Markus E., for example, a 75-year-old man from the settlement of Nutaarmiut in the central part of Upernavik district, has, until recently, hunted since he was a young child, having caught his first seal while out on the ice with his father when he was eight years old. In 2015, Markus admitted to himself that it had been the hardest year of his life, not because of changes in the weather or because of hunting regulations, but because his eyesight was failing. He decided to give his ten sledge dogs to his nephew because he could not look after them any longer. Born on the other side of the island of Nutaarmiut, he moved to Ikerasaarsuk, a small settlement across a narrow isthmus from the village of Nutaarmiut, with his family when he was a child. In 1956, they moved a short way to Nutaarmiut when his father built a house there. Although Markus lamented the end of his hunting days as a result of his failing sight, he was already beginning to find that being a hunter in his seventies was becoming increasingly hard. He was also witness to a hunting way of life that, over the past 20 years or so, has been changing and slowly disappearing. I first visited Nutaarmiut in the late 1980s and it was an exclusively hunting community of around 50 people. Two or three hunters still used kayaks to pursue seals and did not own motorboats. At that time, the settlement had been experiencing a decline in population (today there are around 30 full-time, year-round residents compared to around 75 in 1980, for instance). By the early to mid-1990s, however, most hunters began to fish for Greenland halibut and only a few continued to hunt full-time for a living. The settlement's location near prime fishing grounds for Greenland halibut has meant it has become a centre for landing the catch – a fish-processing plant was built at Nutaarmiut a few years ago and the facility includes several houses for employees, seasonal workers and fishers from other parts of the district. When I spoke to him in June 2015, Markus said he had noticed a lot of seals in the area around the village in recent years – there were fewer hunters now, he said, and so there are fewer people to catch them. He remarked how he himself, even with his eyes letting him down, did not encounter difficulties in knowing where to hunt for seals – he knows their habits and their routes, he said, how and where they swim as they migrate, and where they fish; and so he sets his nets in winter based on this knowledge and he knows the right places to go seal hunting in summer and autumn, something he felt younger hunters do not know.

Above all, however, Markus emphasised the importance of being in the right kind of relationship with the seals he goes in pursuit of. To know about seals is one thing, he said, but to be aware of how the seal wishes, in the end, to be hunted and to give itself to the right hunter who respects it, wants it and needs it, and who knows how to share it, is the way of the real hunter. People may still hunt, but not as their main occupation. Yet, because people have turned to fishing, he said, they no longer know how to hunt – they do not have the right kind of understanding of seals and other animals to be a strong, capable and knowledgeable hunter (*nakuarsuuvoq*). Markus remarked how seals, for instance, are adapting to, and thwarting, the hunting techniques of those who are no longer real hunters. As one example, he pointed to how seals appear to be avoiding nets – at a time when nets were made of a thicker material and would be more noticeable, the seals would still get caught in them; today they are made from an almost transparent nylon which the seals avoid. It is as if, Markus thought, the seals believe hunters are trying to trick them into the nets – a seal will knowingly entangle itself in a net, he said, if it is set by a real hunter and if there is no attempt to conceal its presence.

## SEISMIC SURVEYS AND ABSTRACT SPACES

Local economies along the northwest coast of Greenland, then, are in transition. But while hunting and fishing activities become increasingly difficult for some, as the very nature of hunting changes, and as glaciers melt, crumble and recede, international energy and mining companies have been investigating Greenland's potential as a significant place for oil and mineral extraction. This has given rise to the emergence of what Gavin Bridge (2004) calls an interest in bonanza geographies, the making of resource spaces in which investment from the extractive industries sector and activities of exploration and development appear influential in acting to transform places into zones of speculation and abstraction. The extractive frontier has become part of contemporary Greenland's political life and the subterranean figures in economic calculations for a prosperous future, just as it does for many other countries anticipating or experiencing resource booms. The far northwest has not escaped this interest in discovering what lies below and within the subsurface. Coastal waters are being surveyed for oil, and fells, hills and mountains are sites of exploration for minerals. While mining companies are considering the prospects of mineral-bearing environments, northern Baffin Bay and Melville Bay have seen prospecting and exploratory activity for oil in recent years.

Excited talk of such places emerging as frontier resource spaces, mineral zones and hydrocarbon provinces fills the offices and meeting rooms of government departments, geologists, industry consultants and entrepreneurs in Nuuk (and, currently, Melville Bay is also being scoped out as a new area for halibut and shrimp trawlers to operate in). Such talk, as I have heard it in the capital, does not consider the human history and contemporary human presence of the region, other than insist that companies carry out environmental impact assessments and social sustainability studies as part of the requirements of the regulatory framework within which extractive industries must operate. Instead, northern Greenland is configured as a lively, dynamic space in which strata, igneous intrusions, oceanic trenches, ridges and resources are waiting to be explored and exploited. Maps of Greenland's geology, minerals and licensing blocks for hydrocarbon exploration hang on the walls of those offices and meeting rooms, identifying and locating resources within abstract spaces and emphasising the significance of bonanza geographies (Nuttall 2017). Reacting to the marking out – as well as the marketing – of such frontier spaces, scientists and conservationists are demanding that wildlife and North Greenland ecosystems should be protected, rather than exploited, especially given scenarios suggesting rapid depletion of sea ice, as the Last Ice Area campaign by the World Wide Fund for Nature (WWF) to designate parts of northern Greenland and the Canadian Eastern Arctic illustrates.[2] For those who live there, and who seek to have their voices heard in discussions about development, conservation and wildlife management, the region has become characterised by uncertainty, as well as subject to a hardened global gaze, in a way they have not really experienced before. International companies may see potential arising from deep below the ground and the subsea floor, but the possibility of extractive industries being developed in northern Greenland makes many local residents anxious.

Although a number of projects are going through planning, application and regulatory phases, and while a couple of mines have been given approval and have started operations, Greenland has not yet experienced the minerals or hydrocarbon bonanza that many politicians and business leaders have hoped for. Instead, resource talk in Greenland tends to be characterised by speculation about resource bonanzas and the anticipation of large-scale and lucrative ventures. There was considerable seismic activity in northern Baffin Bay and Melville Bay in 2012 and 2013. Many local people were concerned about it, but some entrepreneurs and municipal officials also hoped that oil development would bring economic benefit. No seismic activities have taken place since then, however. The low price of crude oil is partly a reason and while this may have cast doubt on some future exploratory ventures off the northwest coast, hydrocarbon development has not completely disappeared from the Greenland government's strategy for non-renewable resource development. Bids for a new licensing round for oil and gas exploration in Baffin Bay were invited in 2017, although none had been received by the time the deadline for applications had passed in December that year. The oil companies have shifted their attention for the time being to thinking about the possibilities of hydrocarbon discoveries in the waters off Northeast Greenland.

In the Melville Bay communities of Savissivik and Kullorsuaq, however, as well as in communities further south in the Upernavik district, hunters worry that the effects of seismic surveys linger, even after several years, and that this is evident in the movement of seals away from traditional hunting areas as well as from observing the restless nature of narwhals. Hunters also emphasise their concerns that the impacts of climate change on their livelihoods have been exacerbated by seismic activities as part of oil exploration campaigns. They also worry that the seismic survey vessels will return one day, and are especially concerned about the possible impacts on narwhals, as well as by tighter management regulations for narwhal hunting (the current annual quota for hunted narwhals in Upernavik and Melville Bay is 81).

Just as many scientists say that declining sea ice conditions will result in declining polar bear populations in the Arctic (Peacock et al. 2011), observations also suggest that narwhals are vulnerable and threatened as the ice disappears (Simmonds and Isaac 2007). Marine biologists argue that their restricted core areas of distribution, their attachment to the same migration routes, their small population size and dietary specialisation, as well as the subsistence hunt, make narwhals particularly sensitive to climate change and its effects on sea ice and the Northern marine environment (for example, Laidre et al. 2008; Nielsen 2009). The exact nature of those effects remains uncertain, however. As sea ice now often forms later than it has usually done, narwhals have been observed to delay their departure from coastal areas leading to concern that some are at greater risk of being trapped in the ice when the sea does begin to freeze in the fjords and bays (Heide-Jørgensen et al. 2012).

The regulated hunt for narwhals in Melville Bay, and further north near Qaanaaq, is an essential part of the annual seasonal hunting round, and of the social and economic fabric of family, household and community life. As with other marine mammal hunting activities, narwhal hunting has an important economic aspect and enduring social and cultural value. Hunters have observed some quite significant changes in narwhal movement and behaviour since the seismic activities were conducted, although attributing such change directly to those activities is difficult given the effects of other environmental changes. In research I have been doing with colleagues from the Greenland Climate Research Centre in Nuuk, we have pointed out how hunters say that narwhals appear, as they put it, 'restless' and 'anxious' in Melville Bay, for instance, with some hunters observing that narwhals stay closer to the coast rather than venturing into open water – some say they feel this could be because the narwhals are 'alarmed by' and 'afraid of' something out at sea (Nuttall 2016, 2017). Hunters are aware of how sensitive narwhals are to the presence of hunters, reinforcing scientific concern about the effects of anthropogenic sound. Local hunting regulations in Melville Bay mean that narwhals can only be hunted from a kayak with hand-held harpoons and given concerns that motor boats and noise from rifle shots create problems for narwhals, hunters have expressed their worries that the noise from seismic vessels and airgun pulses have considerable consequences.

It is difficult to determine if such locally observed changes in narwhal behaviour are caused by environmental conditions, shifting sea ice, overall increased shipping traffic or oil and gas exploration. Most likely it is a combination of factors. While environmental changes in the marine ecosystem, such as thinning and declining sea ice, meltwater runoff from tidal glaciers, changing water temperatures, and changes in the migration and distribution of fish also play their part and possibly affect and influence narwhal movement (Greenland halibut are a key prey species for narwhals), hunters say narwhals are not necessarily agitated by such changes alone. They are also aware of how difficult it is to identify a single cause for the changes in narwhal behaviour. The general consensus among hunters is that there are many observed changes in narwhal movement, sea ice, weather patterns and so on, but that there is a multiplicity of factors at work that contribute to this. As one hunter from Nutaarmiut said to me during a community workshop in June 2015:

> I've been here for many years in Nutaarmiut. This year it was really hard to catch halibut. Initially, I thought that narwhals must cause this but because we haven't seen narwhals for some time it must be climate change.

Hunters wish to be included in future work and monitoring of effects of exploratory activities on narwhal hunts (Nuttall et al. 2015; Nuttall 2016), and they point out that there should be greater collaboration between them, government bodies and management institutions, and the oil companies, especially during narwhal census activities. As another hunter from Nutaarmiut put it in summer 2015,

> The most important thing is to have consultation. The oil companies have to come and talk with the hunters in the settlements. It is really

important to combine knowledge before *anything* starts. It is important to develop the respect and relationships right from the beginning.

## CONCLUSIONS

A warming climate is not always what people in Northwest Greenland talk about when they reflect on how their livelihoods and communities have been transforming over the last couple of decades. They are certainly not unaffected by an overheating world, but the effects of climate change on livelihoods which are dependent on the marine environment must be understood in a much wider context of other forms of social and economic change, as well as the influence of national politics and governance frameworks. Everyday life in northern Greenland, a region which, to some extent, is still remote and isolated from Nuuk and other large places on the west coast, not just spatially but in political, economic and cultural senses too (and bad weather frustrates and delays the air link between Upernavik and Ilulissat regularly), is nonetheless circumscribed by the institutions of wider Greenlandic society and the regulations and quotas that are implemented by government bodies for the management of living marine resource use. People in the Upernavik district often complain that they feel disadvantaged – even neglected – because investment appears to be channelled into a few larger towns further south and that decision-making powers about animals and fish are vested in government departments filled with civil servants who do not understand human-environment relations in small communities. Greenland's municipalities were reorganised in 2008, reducing the number from eighteen to four. Upernavik municipality became part of the larger Qaasuitsup municipality, which covered an enormous part of northern Greenland, and decision-making was placed in Ilulissat, further south in the Disko Bay region, rather than in Upernavik. In January 2018, Qaasuitsup was divided into the Avannaata and Qeqertalik municipalities, with the Upernavik region and the rest of Northwest Greenland constituting the bulk of the former municipal area. Many people I know in Upernavik tell me they feel nothing has changed with this administrative reorganisation – the municipality's central offices remain in Ilulissat. During a visit I made north in spring 2018, several friends informed me they did not feel hopeful that economic policy would shift in the direction of Upernavik town and the district's villages.

A range of social, discursive, economic and political practices are at play through which Greenland's environment and resources are given different meanings and emerge as contested and social, political and ideological sites entangled with wider global processes (Nuttall 2017;

Dodds and Nuttall 2019). All this often combines with changes in environment and climate to influence, determine, affect and impact the abilities of people to travel and move around their localities in order to access resources, engage with the non-human entities around them, and hunt and fish the species which sustain them. Currently, it seems that an ability to procure living marine and terrestrial resources – as well as opportunities to sell the products from hunting and fishing – will remain one foundation for the continuity of life in Northwest Greenland's small communities. However, communities in Northwest Greenland are not only challenged by climate change, but by governance systems and institutions that often inhibit and constrain locally specific, long-term resource availability around communities and the rights of individuals to access those resources. Actions for sustainability in an overheating Greenland, and for the survival of the small communities in the far north of the country, must be informed by an understanding of how people in those communities not only experience a complexity of change and perceive the global processes affecting their lives, but how their understandings of human-environment relations arise from being in the world with a multiplicity of human and other than human selves.

## ACKNOWLEDGEMENTS

This chapter draws from research carried out in Northwest Greenland under the auspices of the Climate and Society research programme at the Greenland Climate Research Centre, as well as the EU FP7-funded ICE-ARC (Ice, Climate and Economics – Arctic Research on Change) project and the ArcticChallenge project funded by the Norwegian Research Council.

## NOTES

1. http://nsidc.org/arcticseaicenews/2018/03/arctic-sea-ice-maximum-second-lowest/ (accessed 23 March 2018).
2. Information about WWF's Last Ice Area initiative can be found at: www.wwf.ca/conservation/arctic/lia/ (accessed 23 March 2018).

## REFERENCES

ACIA. 2005. *Arctic Climate Impact Assessment: Scientific Report.* Cambridge: Cambridge University Press.
AMAP. 2012. *Arctic Climate Issues 2011: Changes in Arctic Snow, Water, Ice and Permafrost.* Oslo: Arctic Monitoring and Assessment Programme.

Box, J.E. and D.T. Decker. 2011. Greenland Marine-terminating Glacier Area Changes: 2000–2010. *Annals of Glaciology* 52(59): 91–8.

Bridge, G. 2004. Mapping the Bonanza: Geographies of Mining Investment in an Era of Neoliberal Reform. *The Professional Geographer* 56(3): 406–21.

Briggs, J. 1991. Expecting the Unexpected: Canadian Inuit Training for an Experimental Lifestyle. *Ethos: Journal of the Society for Psychological Anthropology* 19(3): 259–87.

Di, Q., L. Chen, B. Chen, Z. Gao, W. Zhong, R.A. Feely, L.G. Anderson, H. Sun, J. Chen, M. Chen, L. Zhan, Y. Zhang and W.-J. Cai. 2017. Increase in Acidifying Water in the Western Arctic Ocean. *Nature Climate Change* 7: 195–9.

Dodds, K. and M. Nuttall. 2019. Geo-assembling Narratives of Sustainability in Greenland. In U.P. Gad and J. Strandsbjerg (eds), *The Politics of Sustainability in the Arctic: Reconfiguring Identity, Space, and Time*. London and New York: Routledge, pp. 224–41.

Hansen, K. 2008. *Nuussuarmiut: Hunting Families on the Big Headland*. Meddelelser om Grønland 35. Copenhagen: Commission for Scientific Research in Greenland.

Heide-Jørgensen, M.P., R.G. Hansen, K. Westdal, R.R. Reeves and A. Mosbech. 2012. Narwhals and Seismic Exploration: Is Seismic Noise Increasing the Risk of Ice Entrapments? *Biological Conservation* 158: 50–4.

IPCC. 2013. *Climate Change 2013: The Physical Science Basis. Contribution of Working Group I to the Fifth Assessment Report of the Intergovernmental Panel on Climate Change*. Cambridge: Cambridge University Press.

—— 2014. *Fifth Assessment Report. Working Group I: The Physical Science Basis and Working Group II: Impacts, Adaptations and Vulnerability*. Cambridge: Cambridge University Press.

Laidre, K.L., I. Stirling, L.F. Lowry, O. Wiig, M.P. Heide-Jørgensen and S.H. Ferguson. 2008. Quantifying the Sensitivity of Arctic Marine Mammals to Climate-induced Habitat Change. *Ecological Applications* 18: S97–S125.

Lind, S., R.B. Ingvaldsen and T. Furevik. 2018. Arctic Warming Hotspot in the Northern Barents Sea Linked to Declining Sea-ice Import. *Nature Climate Change* 8: 634–9.

Nielsen, M.R. 2009. Is Climate Change Causing the Increasing Narwhal (*Monodon monoceros*) Catches in Smith Sound, Greenland? *Polar Research* 28(2): 238–45.

Nuttall, M. 1992. *Arctic Homeland: Kinship, Community and Development in Northwest Greenland*. Toronto: University of Toronto Press.

—— 2009. Living in a World of Movement: Human Resilience to Environmental Instability in Greenland. In S.A. Crate and M. Nuttall (eds), *Anthropology and Climate Change: From Encounters to Actions*. Walnut Creek, CA: Left Coast Press, pp. 292–310.

—— 2016. Narwhal Hunters, Seismic Surveys and the Middle Ice: Monitoring Environmental Change in Greenland's Melville Bay. In S.A. Crate and M. Nuttall (eds), *Anthropology and Climate Change: From Actions to Transformations*. London and New York: Routledge, pp. 354–72.

—— 2017. *Climate, Society and Subsurface Politics in Greenland: Under the Great Ice*. London and New York: Routledge.

—— 2018. Arctic Weather Words. *Anthropology News* 59(2): 6–10.

Nuttall, M., M. Simon and K. Zinglersen. 2015. *Possible Effects of Seismic Activities on the Narwhal Hunt in Melville Bay, Northwest Greenland*. Unpublished report. Nuuk: Greenland Climate Research Centre/Greenland Institute of Natural Resources.

Peacock, E., A.E. Derocher, G.W. Thiemann and I. Stirling. 2011. Conservation and Management of Canada's Polar Bears (*Ursus maritimus*) in a Changing Arctic. *Canadian Journal of Zoology* 89: 371–85.

Petersen, R. 2003. *Settlements, Kinship and Hunting Grounds in Traditional Greenland*. Meddelelser om Grønland 27. Copenhagen: Danish Polar Centre.

Screen, J.A. and D. Williamson. 2017. Ice-free Arctic at 1.5°C? *Nature Climate Change* 7: 230–1.

Simmonds, M.P. and S. Isaac. 2007. The Impacts of Climate Change on Marine Mammals: Early Signs of Significant Problems. *Oryx* 41(1): 19–26.

Stroeve, J.C., M.C. Serreze, M.M. Holland, J.E. Kay, J. Malanik and A.P. Barrett. 2012. The Arctic's Rapidly Shrinking Sea Ice Cover: A Research Synthesis. *Climatic Change* 110: 1005–27.

Wadhams, P. 2012. Arctic Sea-ice Cover, Ice Thickness and Tipping Points. *Ambio* 41(1): 23–33.

Wang, M. and J.E. Overland. 2009. A Sea-ice Free Summer Arctic Within 30 years? *Geophysical Research Letters* 36, L07502. doi:10.1029/2009GL037820.

Wassmann, P. and T. Lenton. 2012. Arctic Tipping Points in an Earth System Perspective. *Ambio* 41(1): 1–9.

Wilkinson, J. and J. Stroeve. 2018. Polar Sea Ice as a Barometer and Driver of Change. In M. Nuttall, T.R. Christensen and M.J. Siegert (eds), *The Routledge Handbook of the Polar Regions*. London and New York: Routledge, pp. 176–84.

# 5. Volatility
## Understanding Global Capitalism and Climate Change Vulnerability in Mongolia

*Andrei Marin*

INTRODUCTION

In the winter of 2009–10, extremely cold temperatures and heavy snowfall over most of Mongolia's territory killed more than 8 million livestock; 44,000 pastoralist families lost all their livestock, and 164,000 families lost more than half of theirs (Fernandez-Gimenez et al. 2012). This extreme weather event (named *dzud* in Mongolian) severely affected almost one-third of the country's population. In its aftermath, various analysts – from national policy-makers to international donors and humanitarian organisations – have pointed to how pastoralism as a livelihood is increasingly vulnerable to climate change. However, as this chapter will demonstrate, this is not really the case. I will show that the most important cause of vulnerability is not climate change itself, but rather the current economic system that magnifies the vagaries brought about by climate change.

The case of Mongolian pastoralism is potentially illustrative of general mechanisms through which people become vulnerable to so-called 'natural disasters'. In Mongolia, agriculture (almost exclusively pastoralism) contributes significantly to the employment rate (roughly 33 per cent of the active population), and the GDP (13 per cent – the second largest contributor, after mining). The most significant livestock products today are meat (especially mutton) and cashmere. Cashmere, or goat down, has been exploited commercially in Mongolia only since the mid-1970s, but Mongolia is the second biggest producer worldwide, after China. Pastoralism emerged as a dominant livelihood in Mongolia at least 3,000 years ago and has gone through numerous changes and adaptations. Nevertheless, the changes Mongolian pastoralism has witnessed during the past 25 years seem to challenge many of these long-term adaptations. These recent changes relate to the abrupt economic and political reorientation from a long-standing (1924–90) single-party socialist economy to parliamentary democracy and a market economy.

Similar studies show, for instance, that grape-growers in Canada become vulnerable by being exposed simultaneously to cold and wet growing seasons (reducing the quality of the wine) and trade liberalisation (removing trade duties and allowing foreign wines to outcompete the local ones) (Belliveau et al. 2006). Vegetable farmers in Mexico are also negatively affected by market uncertainty and price volatility, despite using irrigation to cope with increasing droughts (Eakin 2003). In Central California, farmers have been affected by the recent financial crisis, which impacted credit markets, local economies and house prices and overlapped with the prolonged drought of 2006–10 (Leichenko et al. 2010). Westerhoff and Smit (2009) identified food insecurity, water scarcity and financial insecurity as the multiple dynamic stressors that exacerbate the effects of Ghanaian farmers' exposure to a changing climate. The present study illustrates a similar situation for Mongolian pastoralists. Their incorporation, through political and economic globalisation, in a volatile system upon which they have almost no influence, dramatically magnifies the risks inherent in their livelihoods.

## PASTORALISM AND CLIMATE CHANGE IN SOUTHERN MONGOLIA[1]

Ondorshil district is a relatively small (roughly 5,000 km²) district in the province of Dundgovi, whose land is 98 per cent covered with arid and semi-arid pastures (Figure 5.1).

At the last census, in 2010, there were 282 herding families in the district, owning on average 260 heads of livestock (camels, horses, cows, sheep and goats), varying between 6 and 1,523 heads/household. The poorer families own mostly sheep and goats (small stock or *bog mal*), while the richer families also own camels, horses and cows (large stock

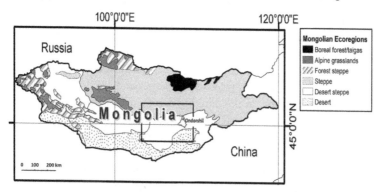

*Figure 5.1* Mongolia and its ecoregions (study area marked in the rectangle)

or *bod mal*). Pastureland is administered by the local government and is legally state property except the land upon which the spring and winter campsites are located (typically 100–200 m$^2$), which are either leased or given for possession to individual herder households by district governments. In Ondorshil, there are 200 such campsites allocated. Different parts of pastureland are used during the four clearly delimited seasons of the extreme Mongolian continental climate (summer in June–August and winter in December–February). While there are guideline schedules for when seasonal pastures should be vacated or used, the decisions about when and where to move are largely informal within each district's borders. Water for human and animal consumption is provided by the 146 wells available, seasonal streams (running mostly in rainy summers) and two salt lakes.

During the last decades, this district, as well as the rest of Mongolia, has been increasingly affected by climate change. At the national level, extreme climatic events (droughts, sand storms, winter disasters) have become more frequent and severe (AIACC 2006). In addition, gradual changes have also been documented, with an increase in average temperature of 1.5°C during the last 60 years (Natsagdorj 2000: 28), and the decade 1990–2000 being the hottest in the last 1,700 years (D'Arrigo et al. 2001). The pastoralists themselves identify more subtle changes in precipitations: the reduction in the spatial extent of rains, that they are more intense and arrive later in the growing season (Marin 2010). Their main adaptation to these changes is to increase the frequency and extent of their migrations in search of good pastures (Marin 2008).

## SOCIO-ECONOMIC CHANGES SINCE 1990

The changes outlined above overlap since 1990 with a rapid political and economic change from a single-party socialist system and command economy to parliamentary democracy and a market economy. This was achieved via a series of 'shock therapy' reforms that included the wholesale privatisation of the government-owned assets, companies and cooperative farms (*negdel*), price liberalisation and elimination of state subsidies, privatisation of the banking sector, tight credit policies and the elimination of trade restrictions. These socio-economic reforms are hinged on the neoliberal principles of state retreat, efficiency, fiscal discipline and trade liberalisation. These principles were strongly argued for in Mongolia by powerful international institutions such as the International Monetary Fund (IMF), the World Bank (WB), the Asian Development Bank (ADB) and the World Trade Organization (WTO).

Although debated among political parties during the early 1990s, the state adhered to these conditionalities and principles.

The result was a de facto transfer of the risk inherent in the pastoralist livelihood from the collective farm and the nationally coordinated agricultural sector, to the individual herder. The Mongolian state has stopped providing many of the services that reduced risk during the communist period: subsidised transportation of people and livestock, maintenance of wells and winter shelters, veterinary services and fodder and hay provision during winter disasters (*dzud*). Importantly, the system of coordinating herders' migration, seasonal pasture rotation, and equal distribution of grazing pressure has also collapsed with the dismantling of the herders' cooperatives (*negdel*).

Finally, an important change was also the price liberalisation of important commodities that included fuel, flour and livestock products such as meat, milk and cashmere (goat down). The speed and intensity of the price liberalisation and the opening up to global commodity markets was extremely swift. As early as January 1991, the Mongolian government adopted the so-called Resolution 20, which effectively removed most price controls for most commodities, and by September 1991, the development plan of the government, which was called the 'Memorandum on economic reform and medium-term policies', was officially adopted. According to Ueno (2000: 8), the Memorandum was 'a joint product of the government and the IMF'. In fact, Ueno more than hints that the adoption of the Memorandum in September 1991 was a precondition for the IMF to approve its stand-by arrangement for a loan one month later (Ueno 2000).

Hence, starting in the early 1990s, the Mongolian herders were abruptly exposed to the volatilities of global commodity markets. Many of them reacted by increasing their herds of goats, since cashmere represents the bulk of yearly household income – approximately 60 per cent (Mercy Corps 2003), hoping to profit from the prices of cashmere, which started to rise on the world market around 1993, and thus enable them to finance their increasingly expensive migrations. They soon found out that the price for cashmere was marred by booms and busts, and that this made profits unreliable, and curtailed their ability to plan migrations and investment in livestock.

This situation is, in fact, not unusual. Empirical evidence shows that trade liberalisation in developing countries and the former communist bloc of Eastern European countries 'translates into a growing feeling of insecurity and uncertainty towards future poverty dynamics' (Montalbano et al. 2007: 204), largely through a multifaceted mechanism of market volatility.

In Mongolia, this volatility and its consequences for poverty are also becoming visible. In order to buffer the volatility of the cashmere market, and since incomes from cashmere are seasonal (occurring in March–May) while migrations occur throughout the year, herders have increasingly started taking commercial loans in order to finance mobility. Not all herders are eligible for loans. While demands for collateral have varied during the past ten years, loans are often given only to the wealthier herders (with at least 200 heads of livestock) and often have prohibitive interest rates (Marin 2008). Worse still, these loans are not predictable. In 2008, as a result of the global financial crisis and the lack of liquidity, Mongolian banks effectively stopped commercial loans to herders. In the same year, China also decided to stop all imports of cashmere (and other commodities), leading to purchase prices in Mongolia plummeting by more than 50 per cent. Herders could neither repay the loans already taken nor access new loans. This situation illustrates how the most important climate change adaptation of Mongolian pastoralists, namely, their mobility, is effectively removed by a set of debilitating circumstances brought about by globalisation.

During the summer of 2006, the drought spell that started in 2004 had reached its peak in southern Mongolia, and in the Ondorshil district this led 80 per cent of the resident families to migrate outside their traditional territories (*nutag*).[2] Many of the pastoralists considered they had never before moved so far away. The families that did not move away split their herds and sent most of their horses and cattle, which are more vulnerable than sheep and goats, to better pastures. Even when horses were kept, they were too weak to be ridden for herding the remaining flocks of sheep and goats, and motorcycles and cars had to be used instead. Using a motorbike for herding would consume 20–30 litres of petrol weekly, the equivalent of the price of a good quality sheep, leading some of the herders to reflect that 'We sell our animals to pay for petrol, in the end' (herder Batsukh).

Migrations can be done either by herding animals on the hoof or by transporting them by truck. If animals are herded on hoof, areas with good pasture and water have to be identified on the way, requiring zigzagging between locations, prolonging the stressful situation for animals and increasing expenses for humans. Even a relatively short 100 km movement took one family eight days and 300 litres of petrol. The alternative is to transport all animals, people and belongings by truck. While this method is swifter and easier, it often leads to animals dying of stress and entails high expenses, as herders rent the truck and driver in addition to paying for the petrol.

Due to the long distance between rural areas and the main markets of the capital city,[3] necessary commodities like livestock vaccines, hay and feed are often much more expensive in the countryside. Conversely, many herders sell their products (mostly meat and cashmere) locally, at prices 20–30 per cent cheaper than in the markets of the capital city, increasing significantly the burden on the household budgets (Marin 2008). In addition to herders' income relying heavily on cashmere, income is also very seasonal, being restricted to March–May, when goats moult and release their cashmere undercoats. The earnings from cashmere sales are typically used to pay back commercial loans taken to finance migration during the preceding summer and autumn.

While some families borrow money from friends and some don't take loans at all, the vast majority of herders increasingly take commercial loans, especially during the recent bad years (Marin 2008). The reliance on commercial loans is surprising given their unfavourable terms for the herders. Interest rates in 2007 were 35 per cent per year (3 per cent per month). The typical herder loans were 500,000 to 1 million *tugrik*[4] for one year. If livestock were used as collateral, only herders with at least 75 small animals (sheep and goats) could get a loan, effectively excluding poorer herders. If other property is used (cars, tents or winter shelter), loans are more readily given for up to 60 per cent of the estimated value. For livestock, one can only borrow a fraction of the livestock market value: maximum US$1/sheep or goat, US$6 for a cow, US$15 for a horse. By 2012, the interest rate had gone down to 2 per cent per month, but the collateral needed was at least 150 *bog mal*. Despite this, the numbers of loans and the amounts borrowed had increased dramatically – it was normal that herders would borrow at least 1 million tugrik *per month*. By 2016, the trend was confirmed but the banks lent only 40 per cent of the collateral value, and increased the demands on collateral. In Ondorshil, one bank claimed to 'own' 80 per cent of the livestock in the district due to herders' defaults on the loans due that year.

Importantly, although the cashmere price has been rising, it has been increasingly volatile, making it an unreliable asset in planning for adaptation. Variations in the price paid by garment producers to cashmere spinning factories have frequently been as high as 200 per cent between the boom and bust years (Figure 5.2).

The fluctuations in the prices paid to Mongolian herders by middlemen who buy largely for spinning companies have registered even more dramatic variations. In Ondorshil, these prices varied from US$42/kg in 2007 to US$12/kg in 2008, to US$55 in 2011 and US$31 in 2012 (Figure 5.2). Moreover, there are great variations within each year between the beginning of the season and its end. The rush to sell first is countered by

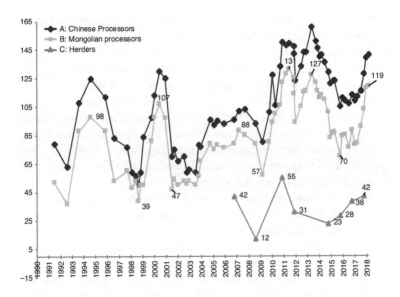

*Figure 5.2* Booms and busts of the cashmere price in US$, illustrated by the price paid by factories to primary processors in China (A) and Mongolia (B), and by the price paid by middlemen to herders in Mongolia (C).

Source: www.gschneider.com for A and B (accessed 26 March 2019) and interviews with herders for C.

the lack of skilled labour for combing the goats. One of my informants did not have time to comb in late March when the price was US$40/kg. By the time he managed to do so, in mid-April, the price had gone down to US$25/kg. This dramatic volatility of the cashmere market can, in my opinion, only be explained in light of an encompassing process of globalisation at work in Mongolia.

## GLOBALISATION IN MONGOLIA

The analysis presented herein relies on approaches that treat globalisation as a sociological process that connects scales (Sassen 2007). Globalisation in this analysis is not a general global-scale transformation towards the integration of economic, political and cultural aspects across nations (Held et al. 1999). Rather than this planetary diffuse condition of benign connectivity, globalisation is better understood as a normative convergence of regulatory frameworks. This is a convergence towards certain economic and political standards and ideals that currently spell out neoliberal capitalism (Sassen 2007; Keskitalo 2008; Nissanke and Thorbecke 2008).

In Mongolia, globalisation as convergence is manifested in two major ways. On the one hand, the most obvious relevant dimension is *economic globalisation*, which relies on increased international trade, direct foreign investment and global financial flows (Keskitalo 2008). On the other hand, but equally important, is the process by which the increased power of international trade conditionalities has often reduced the role and capacity of the national state to regulate its own economy (Budd 1998). This process is an aspect of *political globalisation*, which entails subscribing to the globalisation process through new legal and financial regulations, new regional policies, technologies and industrial developments (Brenner 1999; Keskitalo 2008). Political globalisation, then, can be conceived as the intermingling of layers of power and interests at national and international levels (Shelley 2000). It also indicates that in the present capitalist system, the global scale, with its specific structures and patterns, is being reproduced *within* the national state, enabling a partial denationalisation of the national state, which adheres to international sets of regulatory principles, and (re)produces certain political and economic circumstances that are pushed for by international actors (Sassen 2007). In Mongolia, as I show below, these circumstances pushed various sets of regulatory principles in the past 28 years: from those of the shock therapy in the early 1990s, to the trade regulations inherent in the WTO ascension in 1997 and the recent (2017) IMF loan.

In order to understand the economic globalisation at work in Mongolia, the case of the cashmere market and in particular the role played by China is illustrative. China is the main producer of raw and processed cashmere in the world, today accounting for about 70 per cent of global production (Gankhuyag 2009). With the economic liberalisation of the mid-1980s, Chinese herders were allowed to sell some of their cashmere on the open market, attracting increased interest from Western companies. As previous trade restrictions were gradually removed, competition drove the cashmere prices up by as much as 50 per cent in 1990 (Purcell 1996). Nevertheless, occasional downturns in the economies of important buyers and processors of cashmere, such as the Japanese credit crunch of 1997–98, led to significant drops in the price for cashmere (Mead 1999; see also Figure 5.2). Chinese processors and herders reacted at the time by stockpiling their raw cashmere and avoiding selling for a bad price (Mead 1999). This strategy is possible in China since processors and producers alike benefit from subsidised loans (at 5–7 per cent interest per year; see WB 2003) and other financial incentives that allow them to prolong their loans and likely make a profit when the prices eventually pick up. It is, however, inconceivable for Mongolian herders to hold on to the cashmere for another year if they have a loan at 35 per cent interest per year to pay back. Thus, the touted benefits of economic globalisation have not materialised for Mongolian herders.

## POLITICAL GLOBALISATION AND ITS ECONOMIC CONNECTIONS

Whereas the processes of change in cashmere markets pertain to the economic side of globalisation, there are nevertheless political processes that shape the economic circumstances and in turn affect the vulnerability of different people, sectors and regions. Globalisation is often connected to state retrenchment (Eakin and Lemos 2010). The state may thus devolve some of its administrative and decision-making powers to the local level, or can be pushed by economic and/or political interests to leave the regulation of a certain sector to market forces (Keskitalo 2008). With the privatisation of the early 1990s and the implementation of a fully-fledged neoliberal model of development, all support previously provided by the socialist state to herders was removed, and herders were left to fend for themselves. In two years, between 1990 and 1992, government expenditures were reduced by 50 per cent, by 1993 per capita income was reduced five times[5] and the rate of inflation rose to 325 per cent (Griffin 1995).

This retreat of the state is part of a model of development embraced by Mongolian politicians and pushed by important international donors since the beginning of the transition period. The reform packages which constituted the conditionalities for loans from international donors focused on economic growth and export-growth industries without addressing the issue of poverty and inequality in income distribution (Rossabi 2005). The effects of the neoliberal take on development are thus reflected in the reduction of economic safety nets, a common effect of budget discipline policies advocated by international financial institutions worldwide (Stiglitz 2002).

It is important to understand the intermingling of the political and the economic aspects of globalisation. With the neoliberal reform, for example, there were certain conditions imposed, particularly regarding Mongolia's opening to international trade. With the accession to the WTO in 1997, Mongolia was forced to renounce its policy of discouraging exports of raw cashmere and effectively remove the export tariffs imposed (US$4/kg) by 2007. This resulted in domestic processors losing much of their raw materials, as most of the cashmere is since exported to China (Marin 2008). In 1994, following a government ban on cashmere export, 50 new cashmere mills appeared in Mongolia (Mead 1999). Today, however, following a decade of free trade, the only ones who have profited are middlemen who smuggle cashmere over the border to China.

The credit restrictions and high interest rates in Mongolia stem from the tight monetary policies implemented as part of the neoliberal reform package. In Mongolia, the IMF has encouraged prohibitively high interest rates for commercial loans since the early 1990s, as a strategy against an alleged high risk of default on the loans (Griffin 2003). Nevertheless, this seems to be contradicted by reports from banks that the rate of default was a small fraction of 1 per cent (Griffin 2003). My empirical findings also indicate that a low default rate was still the case for herders in the period 2006–12. It was only in 2016 that a loan manager in Ondorshil expressed concern about high default rates.

An alternative explanation for the high interest rates policy seems to be that there has been a lack of competition in the banking sector (Griffin 2003). Foreign aid may have acted as a substitute for both domestic savings and a badly needed tax reform leading to diminished revenues and an effective weakening of the state (Griffin 2003). In addition, large inflows of liquidities in the form of development aid have led to the appreciation of the national currency, further reducing the ability of domestic industries to compete against foreign products on the international markets, a pattern increasingly acknowledged by critics of development aid (for example, Moyo 2010).

The recent economic changes in Mongolia have also been driven by (geo-) political interests, decisions and ideologies. First, Mongolia's geographical position between Russia and China is undoubtedly an element in the interest shown by large international donors in promoting and supporting the neoliberal market-driven reforms. Second, the political lobbies in the Western donor countries have pushed a set of loan conditionalities that would favour the interests of Western companies in Mongolia, especially in the mining and energy sectors. For example, in 1999, when the Mongolian state wanted to privatise the power plant supplying 70 per cent of the heat and energy to the capital city, intense American lobbying by the US Assistant Secretary of State and Vice President Al Gore tried (unsuccessfully) to press the Mongolian government to sell to a US energy company. In 2008, the Mongolian government's decision to nationalise the Tavan Tolgoi coal mine prompted a campaign to stop US foreign aid to Mongolia because the nationalisation went against the principles of the free market. The campaign, reminiscent of Cold War rhetoric, presented Mongolia as a prospective enemy of the West, potentially fostering the interests of Russia and China.

One of the main characteristics of globalisation is its supraterritorialisation, whereby borders and other territorial rules no longer matter as far as the flow of goods and services is concerned (Scholte 2000). Others refer to this as a time-space compression (Harvey 1989), leading to a world that is speeding up and spreading out, although not in the same ways for all of us (Massey 1994). This compression is usually said to cause the reduction of power of the national state (Brenner 1999; Sassen 2007; Swyngedouw 2007). Yet, some scholars have convincingly argued that globalisation may also empower an individual state's citizens by facilitating their access to transnational advocacy networks (Appadurai 2003; Milanovic 2003). Political globalisation, therefore, may manifest itself in the reduced power of the national state, and/or increased power of international civil society. As Sassen (2007) points out, although globalisation entails a partial embeddedness of global economic and political standards and ideals into national contexts, the relation often involves *mutual* imbrications, and therefore we need to understand the particular ways different countries negotiate and institutionalise these standards and ideals, as well as capitalism more generally.

The following example illustrates this point. The ways that Mongolia and Nepal have engaged with the political-economic apparatus of the WTO have been very different. Nepal's ten-year negotiation to join the WTO involved a regional civil society organisation (South Asia Watch on Trade, Economics and Environment – SAWTEE) and an extensive network of national non-governmental organisations (NGOs), which allowed the government to secure particularly favourable terms of trade for sensitive agricultural products: a 51 per cent tariff, to be reduced to 42 per cent after the transition period (Rajkarnikar 2005). The Mongolian story of accession to the WTO is very different. During the six years of negotiation for WTO accession (1991–97), the Mongolian government has monopolised the negotiation process, without involving civil society or representatives of the most affected industries (Tsogtbaatar 2005). The cashmere industry was the most 'cosmopolitan' industry, having been involved with Western trading partners even during the communist regime (Songwe and Mangvan 2001), and the most vocal in challenging the Mongolian government on the consequences of the WTO accession (Tsogtbaatar 2005). Removal of trade barriers would mean that much of the raw cashmere could leave the country, forcing the Mongolian textile industry to import yarn. These worries have proven justified: when Mongolia joined the WTO in 1997, it applied a 30 per cent tariff on the sale price of cashmere. After 1997, the tariff was turned into a flat export tax of US$6/kg cashmere, applied ever since. Had the 30 per cent been kept in 2018 it would have led to an export tax at least three times as high. In practice, this WTO conditionality made it cheaper to export raw cashmere. Market economic thinking would argue that this removal of trade barriers made it easier for herders to take advantage of selling cashmere on the international markets, with higher profit margins. In reality, it led to a precarious situation, whereby climate change-induced unpredictability was turned into fully-fledged risks by the volatile market.

## VOLATILITY: WHEN SPEEDING UP AND SPREADING OUT GO WRONG

The most obvious outcome of globalisation for the Mongolian herders is the cashmere price volatility. China is the most important actor in the cashmere market, effectively deciding world prices for both raw and processed cashmere (Marin 2008). The Mongolian herders, on the other hand, are vulnerable to the purchasing policy of China, in turn influenced by the world demand for luxury cashmere garments. The integration of the Mongolian cashmere market with the international one has undoubtedly contributed to herders getting a better price for their cashmere (Pomfret 2000). Yet, these prices are not predictable and have proven to be a false friend for the herders, failing them when they needed to pay back loans taken to avoid dire climatic circumstances. Had the market been free and equitable as the proponents of neoliberalism would have it, the Mongolian companies would be able to access European, American and Japanese clothing markets directly and would not depend on Chinese yarn to function. If this access were ensured, Mongolian herders would be able to sell cashmere to local processors.

At present, the Mongolian processors' demand for raw cashmere is not met, forcing them to operate at half capacity (Gankhuyag 2009). In 2008, though, there was a low demand from domestic processors. The reason for this is another illustration of volatility, as explained by the National Bank of Mongolia (Damdinsuren 2009). With the global financial crisis of 2008, the price of Mongolian mining products fell, resulting in a foreign trade deficit and reduced foreign currency reserves. The crisis overseas resulted in foreign institutions retracting their loans to Mongolian banks and delaying further proposed investments, while Mongolians overseas reduced their remittances. This was a perfect financial storm that caused a lack of liquidity in the banking sector, and forced banks to stop lending (Sassen 2007).[6] Since China stopped the purchase of raw cashmere the same year, and with domestic processors strapped for cash, prices for raw cashmere plummeted. This situation points to a market that is 'integrated' and 'free' only on a superficial level, allowing for political interference and extreme volatility. Indeed, the Chinese politically controlled trade has been termed a 'half-open' system: open for exports and closed for imports (Hu 2004). This allows political control of the cashmere prices and supports the recurrent critique of globalisation as ad hoc governance without a government (Stiglitz 2002).

These observations illustrate that market integration (and economic globalisation more generally) has the effect of tightly connecting small-scale producers to large-scale political and economic processes over which they often have little influence. This situation has been identified in system-theory language as hyperconnectivity, whereby systems are dependent on flows of information circulating at such high speed that they have little or no buffering capacity: they are systems with 'no room for error', where vulnerability and dependence are reinforced by increasing speed and connectivity (Homer-Dixon 2006: 114). In such systems, damage to one part of a system (for example, a local financial crisis) has cascading effects that propagate rapidly to other parts of the system before anyone can intervene. The effect of the 2008 financial crisis on plummeting prices of cashmere, and the herders' ability to cope with the winter disaster that struck in 2009, are therefore a good example of this dangerous connectivity.

The new systems that are characterised by global hyperconnectivity are affected by risks that are more difficult to predict. In economic parlance, such risks are often referred to in terms of vulnerability emerging from the high volatility of trade openness and terms of trade and have been shown to play a significant role in ex-communist economies (Montalbano et al. 2007; Nissanke and Thorbecke 2008). Still, I propose that there are more fundamental mechanisms of capitalism at play here.

The case of cashmere production in Mongolia illustrates the strength and reach of the idea of risk as an intrinsic part of our economic model. So much so, as Foucault has argued (2002: 66), that we are all encouraged to perceive risk as a normal part of the human condition. In Mongolia, the communist political-economic system echoed the medieval sets of relations that committed or even obliged the 'nobleman' to protect the 'subject', making the former accountable for the wellbeing of the latter. In the present neoliberal system, the state is committed instead, as Foucault puts it (2002: 66), to protecting from risk a set of rules of governance. Such rules, like the WTO conditionalities and the market deregulations illustrated above, are the very rules that make the lives of people like Mongolian pastoralists increasingly risky.

## FAIR TRADE AS A POLANYIAN COUNTERMOVEMENT?

Given the rather discouraging picture portrayed above, one would be forgiven for seeing the most recent developments in the global cashmere markets as a much-needed silver lining. Major media outlets (for example, Mead 1999), have documented the plight of Mongolian herders under climate change, and have often identified cashmere production as a culprit that increases risk by land degradation. Some fashion houses that specialise in cashmere products have recently proposed environmentally friendly and socially just cashmere garments: so-called 'fair cashmere'. The argument is that fair trade in the cashmere industry would give herders higher profit, encouraging them, therefore, to keep fewer goats. This would reduce pasture degradation and help herders avoid weather-related disasters. The cashmere industry itself has thus produced a narrative that buying luxury cashmere garments supports a sustainable alternative to high-speed 'fast fashion', in turn identified as the epitome of unsustainability intrinsic in current capitalism (Brooks 2015). The 'fair' cashmere industry is presented as 'sustainable luxury fashion' that supports 'indigenous cultures' (Gardetti and Rahman 2016). There are nowadays 'sustainable fashion shows' from San Francisco to Seoul, and major actors in the cashmere industry (for example, Loro Piana Corporation) rebrand themselves as defenders of sustainable development.

This switch towards ethically conscious markets can be construed as a Polanyian countermovement (Polanyi 2001) whereby society opposes the logic of the market with a search for fair production and consumption relations anchored in social relations. It would be wrong to do so. In my opinion, the fact that the best 'countermovement' is a fair trade one is illustrative of the lack of alternatives. I agree with Guthman (2007) that fair trade reinforces the market as the only locus of regulation, proving once more that 'neoliberal political economies and subjectivities have delimited the possible' (Guthman 2007: 456). In other words, this is not a countermovement but merely a marketing strategy in a highly competitive industry. It illustrates the spread of a brand of corporate capitalism in which the flipside of risk is the opportunity to target consumers who are concerned with 'fair trade' or 'green consumption'.

In the end, it seems that 'fair cashmere' is illustrative of how 'the logic of capitalism and the logic of resistance against capitalism have converged' (Hickel and Khan 2012: 206).[7] Capitalism shows in this a subtle ability to coopt dissent and rebellion. Swyngedouw (2007) and others (Žižek 1999; Wilson and Swyngedouw 2015) argue that this has been made possible because the present capitalist condition is a post-political one, in which global-scale consensus has replaced any ideological critique of the current system.

Narratives such as 'fair cashmere' and 'sustainable fashion' accept the capitalist mode of production and consumption while rebelling only against its consequences. The new capitalist ethic, then, relies on high-priced commodities that claim to be advancing social justice, though through egotistic consumption (Žižek 2009). What we are left with is 'the new commodity fetishism' in which global capitalism is supposedly challenged *through* consumption, in fact reaffirming the very structures it sets out to challenge (Hickel and Khan 2012: 213). The invisible hand of the market therefore secures its reach, not *in spite* of the language of anti-conformity and consumer ethics of fair trade, but *because* of it. As Comaroff and Comaroff (2001) put it, the invisible hand becomes the Gucci-clad fist that guides political, material and social forms in neoliberal global capitalism.

In the southern part of Mongolia, herders do not perceive any effects of so-called 'fair trade'. They continue to sell to middlemen for the price set by highly volatile markets where prices can vary by 200 per cent from one week to the next, depending on economic and political circumstances beyond their reach. If there were any 'fair trade' middlemen in Ondorshil, they would not be able to reduce volatility for the herders. A slightly better price paid to producers would not address the structural vulnerabilities that are intrinsic to the present system. Since the present neoliberal system is 'shot through with deeply incompatible interests' (Hickel and Khan 2012: 204), reducing vulnerability to risk entails solutions outside the market.

If there is any silver lining, therefore, it should be sought elsewhere, not in the so-called 'fair trade market', but in regulations that protect small-scale producers from the vagaries of global markets. Examples of successful engagements with the global markets can be found in the case of Caribbean farmers' integration with international banana markets, which increased their incomes and food security, and in the case of Ivorian farmers producing cotton for the global market, who also acquired increased stable incomes (Robbins 2006). In these cases and others, success could be attributed to specific institutions that absorbed some of the shocks and volatility resulting from market integration. Institutions such as national pricing boards, producer cooperatives, social networks, and local modes of exchange and systems of credit (Robbins 2006: 214) seem to amount to a more viable countermovement.

## CONCLUSION

The case presented here shows how Mongolian herders' 'vulnerability to climate change' is the result of the volatility deriving from their engagement with the current neoliberal capitalist system. It illustrates how economic and political globalisation bring about new sets of economic, financial and legal regulatory principles and relations. Finally, the analysis proposes that the alternative to a system that exacerbates the risks of climate change is not mere tinkering with market regulations or more socially responsible corporations that rebrand themselves in order to maintain their market share.

Recent debates about the differences between inegalitarian (liberal) and social models of capitalism provide some vision for an alternative. They propose that many of the differences between so-called cut-throat, liberal market capitalism, on the one hand, and 'cuddly', coordinated market economies, on the other, stem from pressures for fast turn-around of labour and capital, and from their promotion (or erosion) of social solidarity (Hall and Soskice 2001; Thelen 2014).

Furthermore, research proposes that at a macro-economic scale, volatility in markets and vulnerability to trade shocks more specifically can be counteracted by specific, forward-looking policies that enhance the coping mechanisms of individual households beyond simple risk insurance schemes (Montalbano et al. 2007).

This suggests two antidotes that may provide a much-needed cooling down of the overheated and increasingly volatile reality of the Mongolian herders. First, recognising that there is no single way to prosperity, Mongolian capitalism as a national-scale development paradigm has to go through a re-evaluation of priorities. This should bring the national socio-economic context away from its current neoliberal incarnation and closer to socially empowering, coordinated forms of market economy that reduce vulnerability to shocks, be they climatic or economic. Second, since this re-evaluation will most likely not happen spontaneously, Mongolians themselves will need to bring about a Polanyian countermovement to the neoliberal version of capitalism today. The young age of the population, high levels of political engagement and increased collaboration of Mongolian civil society with global counterparts are all encouraging signs in this direction.

## NOTES

1. The empirical material presented herein has been collected during the past 12 years in the central and southern part of the country, in fieldwork periods varying between two weeks and 11 months, using participant observations, direct semi-structured interviews and life-history conversations.
2. Migration outside customary territories is termed *otor* and represents an unusual situation that herders usually try to avoid.
3. Mongolia is a country half the size of India (1.5 million km$^2$), with very few paved roads.
4. At the time, this was equivalent to US$450–900 (US$1 was equivalent to approximately 1,150 *tugrik*).

5. Per capita income continued to fall, from US$478 in 1992 to US$374 in 1999 (NSOM 2004).
6. Several authors (Griffin 2003; Reinert 2004; Rossabi 2005) have pointed out that even when banks do offer loans, the interest rates are so high that they preclude many companies from accessing loans and investing in infrastructure and their development.
7. An illustrative example of this, given by Hickel and Khan (2012), is a US$200 cashmere cardigan printed with Che Guevara's face on sale in a major international airport.

## REFERENCES

AIACC. 2006. *Climate Change Vulnerability and Adaptation in the Livestock Sector of Mongolia. A Final Report Submitted to Assessments of Impacts and Adaptations to Climate Change.* Washington, DC: The International START Secretariat.

Appadurai, A. 2003. Grassroots Globalization and the Research Imagination. In A. Appadurai (ed.), *Globalization.* Durham, NC and London: Duke University Press, pp. 1–21.

Belliveau, S., B. Smit and B. Bradshaw. 2006. Multiple Exposures and Dynamic Vulnerability: Evidence from the Grape Industry in the Okanagan Valley, Canada. *Global Environmental Change* 16(4): 364–78.

Brenner, N. 1999. Beyond State-centrism? Space, Territoriality and Geographic Scales in Globalization Studies. *Theory and Society* 28: 39–78.

Brooks, A. 2015. *Clothing Poverty. The Hidden World of Fast Fashion and Secondhand Clothes.* London: Zed Books.

Budd, L. 1998. Territorial Competition and Globalisation: Scylla and Charybdis of European Cities. *Urban Studies* 35: 663–85.

Comaroff, J. and J.L. Comaroff. 2001. Millenial Capitalism: First Thoughts on a Second Coming. In *Millenial Capitalism and the Culture of Neoliberalism.* Durham, NC and London: Duke University Press, pp. 1–56.

D'Arrigo, R., G Jacoby, D. Frank, N. Pederson, E. Cook, B. Buckley, B. Nachin, R. Mijiddorj and C. Dugarjav. 2001. 1738 Years of Mongolian Temperature Variability Inferred from a Tree-ring Width Chronology of Siberian Pine. *Geophysical Research Letters* 28: 543–6.

Damdinsuren, B. 2009. *Asset Price Bubbles and Challenges to Central Banks. The Case of Mongolia.* National Bank of Mongolia. Available at: www.mongolbank.mn/documents/tovhimol/group5/05.pdf (accessed November 2018).

Eakin, H. 2003. The Social Vulnerability of Irrigated Vegetable Farming Households in Central Puebla. *Journal of Environment and Development* 12(4): 414–29.

Eakin, H. and M.C. Lemos. 2010. Institutions and Change: The Challenge of Building Adaptive Capacity in Latin America. *Global Environmental Change – Human and Policy Dimensions* 20(1): 1–3.

Fernandez-Gimenez, M.E., B. Batkhishig and B. Batbuyan. 2012. Cross-boundary and Cross-level Dynamics Increase Vulnerability to Severe Winter Disasters (Dzud) in Mongolia. *Global Environmental Change* 22: 836–51.

Foucault, M. 2002. *Power: The Essential Works of Michel Foucault 1954–1984 (Essential Works of Foucault 3)*, ed. J.D. Faubion. London: Penguin.

Gankhuyag, S.O. 2009. A Study of the Competitive Advantages of Cashmere Industry in Mongolia. National Sun Yat-Se University. Available at: https://bit.ly/2DkLxL0 (accessed November 2018).

Gardetti, M.A. and S. Rahman. 2016. Sustainable Luxury Fashion: A Vehicle for Salvaging and Revaluing Indigenous Culture. In M.A. Gardetti and S.S. Muthu (eds), *Ethnic Fashion, Environmental Footprints and Eco-design of Products and Processes*. Singapore: Springer Science+Business Media, pp. 1–18.

Griffin, K. 1995. *Poverty and the Transition to the Market Economy in Mongolia*. London: Macmillan.

—— 2003. The Macroeconomics of Poverty. In K. Griffin (ed.), *Poverty Reduction in Mongolia*. Canberra: Asia Pacific Press, pp. 1–28.

Guthman, J. 2007. The Polanyian Way? Voluntary Food Labels as Neoliberal Governance. *Antipode* 39(3): 456–78.

Hall, P.A. and D. Soskice. 2001. An Introduction to Varieties of Capitalism. In P.A.

Hall and D. Soskice (eds), *Varieties of Capitalism. The Institutional Foundations of Comparative Advantage*. Oxford: Oxford University Press, pp. 1–45.

Harvey, D. 1989. *The Condition of Post-modernity: An Inquiry into the Origins of Culture Change*. Cambridge: Blackwell.

Held, D., A. McGrew, D. Goldblatt and J. Perraton. 1999. *Global Transformations: Politics, Economics and Culture*. Cambridge: Polity Press.

Hickel, J. and A. Khan. 2012. The Culture of Capitalism and the Crisis of Critique. *Anthropological Quarterly* 85(1): 203–27.

Homer-Dixon, T. 2006. *The Upside of Down. Catastrophe, Creativity, and the Renewal of Civilization*. Washington, DC: Island Press.

Hu, A. 2004. The Free Trade Agreement Policy for Northeast Asian Countries and ASEAN: A View from China. In H. Hirakawa and Y.-H. Kim (eds), *Co-design for a New East Asia after the Crisis*. Tokyo and Berlin: Springer Verlag, pp. 51–66.

Keskitalo, E.C.H. 2008. *Climate Change and Globalization in the Arctic. An Integrated Approach to Vulnerability Assessment*. London: Earthscan.

Leichenko, R., K. O'Brien and W.D. Sollecki. 2010. Climate Change and the Global Financial Crisis: A Case of Double Exposure. *Annals of the Association of American Geographers* 100(4): 963–72.

Marin, A. 2008. Between Cash Cows and Golden Calves. Adaptations of Mongolian Pastoralism in the 'Age of the Market'. *Nomadic Peoples* 12: 75–101.

—— 2010. Riders under Storms: Contributions of Nomadic Herders' Observations to Analysing Climate Change in Mongolia. *Global Environmental Change –Human and Policy Dimensions* 20: 162–76.

Massey, D. 1994. *Space, Place, and Gender*. Minneapolis, MN: University of Minnesota Press.

Mead, R. 1999. Letter from Mongolia: The Crisis in Cashmere. *The New Yorker* 74: 56–64.

Mercy Corps. 2003. Cashmere Market: Processors' Requirements for Raw Cashmere. Gobi Regional Economic Growth Initiative. Available at: https://bit.ly/2PaGXpn (accessed August 2017).

Milanovic, B. 2003. The Two Faces of Globalization: Against Globalization as We Know It. *World Development* 31: 667–83.

Montalbano, P., A. Federici, U. Triziuli and C. Pietrobelli. 2007. Trade Openness and Vulnerability in Central and Eastern Europe. In M. Nissanke and E. Thorbecke (eds), *The Impact of Globalization on the World's Poor: Transmission Mechanisms*. New York: Palgrave Macmillan, pp. 204–34.

Moyo, D. 2010. *Dead Aid: Why Aid is Not Working and How There is a Better Way for Africa*. New York: Farrar, Strauss ancd Giroux.

Natsagdorj, L. 2000. Climate Change. In P. Batima and D. Dagvadorj (eds), *Climate Change and its Impacts in Mongolia*. Ulaanbaatar: JEMR Publishing.

Nissanke, M. and E. Thorbecke. 2008. Globalization and Poverty in Asia. Can Shared Growth be Sustained?. In M. Nissanke and E. Thorbecke (eds), *Globalization and the Poor in Asia. Can Shared Growth be Sustained?* New York: Palgrave Macmillan.

NSOM (National Statistical Office of Mongolia). 2004. *Mongolia in a Market System Statistical Yearbook*. Ulaanbaatar. NSOM.

Polanyi, K. 2001. *The Great Transformation: The Political and Economic Origins of Our Time*. Boston, MA: Beacon Press.

Pomfret, J. 2000. Mongolia Beset by Cashmere Crisis: Herders, Mills Struggle in New Economy. *Washington Post*, 17 July.

Purcell, T. 1996. Scotland and China and Cashmere Trade. Available at: www1.american.edu/projects/mandala/TED/cashmere.htm (accessed November 2015).

Rajkarnikar, P.R. 2005. Nepal: The Role of an NGO in Support of Accession. In P. Gallagher, P. Low and A.L. Stoler (eds), *Managing the Challenges of WTO Participation. 45 Case Studies*. Cambridge: WTO and Cambridge University Press, pp. 420–9.

Reinert, E.S. 2004. Globalisation in the Periphery as a Morgenthau Plan: The Underdevelopment of Mongolia in the 1990s. In E.S. Reinert (ed.), *Globalization, Economic Development and Inequality. An Alternative Perspective*. Cheltenham: Edward Elgar Publishing, pp. 157–205.

Robbins, P. 2006. *Political Ecology. A Critical Introduction*. Oxford: Blackwell.

Rossabi, M. 2005. *Modern Mongolia, from Khans to Commissars to Capitalists*. Berkeley, Los Angeles, CA and London: University of California Press.

Sassen, S. 2007. *A Sociology of Globalization*. New York: W.W. Norton & Company.

Scholte, J.A. 2000. *Globalization: A Critical Introduction*. New York: St. Martin's Press.

Shelley, B. 2000. Political Globalization and the Politics of International Non-governmental Organisations: The Case of Village Democracy in China. *Australian Journal of Political Science* 35: 225–38.

Songwe, V. and B. Mangvan. 2001. *Mongolia Cashmere Trade Policy*. World Bank Report. Available at@ https://bit.ly/2DjpDb5 (accessed February 2017).

Stiglitz, J.E. 2002. Globalism's Discontents. *The American Prospect* 13(1): 1–14.

Swyngedouw, E. 2007. Impossible/Undesirable Sustainability and the Post-political Condition. In D. Gibbs and R. Krueger (eds), *The Sustainable Development Paradox*. New York: Guilford Press, pp. 13–40.

Thelen, K. 2014. *Varieties of Liberalization and the New Politics of Social Solidarity*. New York: Cambridge University Press.

Tsogtbaatar, D. 2005. Mongolia's WTO Accession: Expectations and Realities of WTO Membership. In P. Gallagher, P. Low and A.L. Stoler (eds), *Managing the Challenges of WTO Participation. 45 Case Studies*, Cambridge: WTO and Cambridge University Press, pp. 409–19.

Ueno, H. 2000. Assessing Transition Policies: The Case of Mongolia. *Kobe University Bulletin Paper* 8(1): 127–68. Available at: https://bit.ly/2OnTjVM (accessed February 2017).

WB. 2003. From Goats to Coats: Institutional Reform in Mongolia's Cashmere Sector. The World Bank. Available at: https://bit.ly/2JCu8Oz (accessed November 2018).

Westerhoff, L. and B. Smit. 2009. The Rains are Disappointing Us: Dynamic Vulnerability and Adaptation to Multiple Stressors in the Afram Plains, Ghana. *Mitigation and Adaptation Strategies for Global Change* 14: 317–37.

Wilson J. and E. Swyngedouw. 2015. Seeds of Dystopia: Post-politics and the Return of the Political. In J. Wilson and E. Swyngedouw (eds), *The Post-political and Its Discontents: Spaces of Depoliticisation, Spectres of Radical Politics*. Edinburgh: Edinburgh University Press, pp. 1–22.

Žižek, S. 1999. *The Ticklish Subject – The Absent Centre of Political Ontology*. London: Verso Books.

—— 2009. *First as Tragedy, Then as Farce*. London: Verso Books.

# 6. The Dark Side of Progress
## The Intersections of Climate Change, Neoliberalism and Modernity in Peru

*Astrid B. Stensrud*

INTRODUCTION

In February 2014, Peruvian newspapers reported yet another extreme weather event in the southern Peruvian Andes. In the province of Caylloma, in the region of Arequipa, the authorities calculated that the drought and frost had caused an economic loss of 14 million *soles* (US$4.9 million).[1] 'These economic losses are affecting 5,284 families in the province, and that is why we solicit the declaration of emergency', the secretary of the Civil Defence of the municipality of Caylloma told RPP Noticias (2014). In the highland districts, fodder for llamas, alpacas and sheep was becoming scarce because of the lack of rain, causing the death of 12 per cent of alpaca *crías* (babies). In Colca Valley, ten villages and 6,000 hectares of cultivated areas were affected by the frosts, and 1,059 hectares, growing mostly green beans, potatoes and maize, were lost (RPP Noticias 2014). Therefore, the provincial municipality asked the Regional Government of Arequipa to declare an emergency in Caylloma province, where the main economic activity is small-scale agriculture.[2] Although this was the first serious drought in ten years, extreme weather events have become more frequent in the past decade. Since 2011, the municipal authorities of Caylloma province have declared states of emergency every year, after large quantities of crops have been ruined and thousands of animals killed by heavy rain, hail and snowfall, and irregular frost periods (Peru.com 2011; RPP Noticias 2012; Buho 2013; RPP Noticias 2013). Infrastructure, such as irrigation canals, reservoirs, roads and bridges, has also been ruined by extreme weather. In April 2014, a group of mayors from the highland districts travelled to the capital, Lima, to present their grievances and petitions to the government: financial support, insurance for llamas and alpacas, and agrarian insurance. They did not receive any economic help, however, but were bought off with minor, short-term measures like medicines for the alpacas in the highlands and 2 kg of seeds for each farmer in the valley.

This experience resonates with a structural feeling of abandonment by the state in the Peruvian highlands, where people often refer to themselves and their villages as 'forgotten' by the government (Rasmussen 2015). The Andean highlands have been structurally marginalised throughout the history of the Peruvian Republic: while the coast has been characterised as capitalist, urban, 'white' and 'modern', the highlands have been regarded as poor, 'indigenous' and 'backwards'. This is in part a result of 'uneven development', understood as the process and pattern of the production of nature under capitalism (Smith 1984). Regions are categorised as developed or developing, and reorganised into an historical queue, where temporality is reduced to a singular model of development and only one way forward is possible (Massey 2006; see also Harvey 2006). In Peru, however, this 'uneven development' is reproduced in a racialised political geography, and is thus part of what was earlier called 'internal colonialism' (Gonzalez Casanova 1965; Quijano 2000), and which later has been analysed more broadly in terms of 'coloniality'. Following Quijano (2000), Mignolo (2000) and Escobar (2007), among others, 'coloniality' is not only the heritage of colonialism, but is constitutive of modernity. This implies, for example, identifying the origins of modernity with the Conquest of America and the control of the Atlantic after 1492, and a persistent attention to colonialism and the making of the capitalist world system as constitutive of modernity (Escobar 2007: 184). What Quijano calls the 'coloniality of power' is a matrix of power in which hierarchies are codified according to ideas of race, class and gender (Quijano 2000). These intersectional inequalities are not only experienced, but also constantly reproduced in everyday life, and these hierarchies are today legitimised by 'culture', education and geography (de la Cadena 2000).

Coloniality is today not only evident in global economic inequality and racism, but also in climate change and environmental crisis. The long-term effects of climatic and environmental changes have been denominated 'slow violence' (Nixon 2011). Slow violence occurs gradually and out of sight, a violence of delayed destruction that is dispersed across time and space, an attritional violence that is typically not viewed as violence at all (Nixon 2011). Kirsten Hastrup (2009) argues that climate change is a chronic disaster; it is diffuse and long term, but it alters fundamentally the foundations of people's lives. For the past two decades, climate change has severely affected the livelihoods of indigenous peasants in the Andean highlands. In the same period, the effects of the deregulation of markets and lack of state support and subsidies to agriculture have made it very difficult to survive as a small-scale organic farmer in the highlands.

This chapter contributes to the understanding of the intersections of climate change, modernisation and neoliberalism by analysing ethnography from the Colca Valley in southern Peru. The ethnographic material was generated during 13 months of fieldwork in Colca Valley and Majes in Caylloma province, Arequipa Region, in 2011, 2013 and 2014. My main argument is that the effects of climate change and neoliberal economic policies reinforce each other and adversely affect small-scale agriculture. I will discuss how the impacts of global warming and global markets affect not only water access and livelihoods, but also local concerns over food safety and life quality. The people that are most affected by climate change in Peru are the ones living in high altitudes, and these areas are not only the most climate-sensitive areas, but also the most poverty-stricken and stigmatised in Peru. Programmes of development and adaptation, which are embedded in the ideology of neoliberal modernity, offer technological 'fixes', which may lead to victim blaming and de-politicisation (Ferguson 1994; Li 2007). I will first discuss how neoliberal economic policies have been implemented in Peru, affecting relationships to land and water. Then, I examine how climate changes are experienced by small-scale farmers in Colca, and how they intersect with the effects of free market policies. Finally, I discuss ideas of progress, modernisation and technology in Peru.

## NEOLIBERAL CAPITALISM: DEREGULATIONS OF LAND, WATER AND PRICE CONTROL

Since the Conquest and until the 1969 land reform, the unequal distribution of land was a major problem in Peru, where powerful estate owners also controlled the access to water. Before 1969, only 3.9 per cent of the population controlled 56 per cent of the land (Seligmann 1995). President Velasco's (1968–75) agrarian reform intended to give land to the indigenous peasants and labourers who worked it, and the radical reform ended the dominance of the landowning elite by replacing the large private estates with agrarian cooperatives (Collier 1978). After Velasco's fall from power in 1975, the turn towards neoliberal economic policies started during the regime of President Morales (1975–80). Neoliberalism, according to David Harvey (2005), is a set of policies and practices inspired by a political ideology according to which human wellbeing can best be advanced by liberating individual entrepreneurial freedoms and skills within an institutional framework that secures private property rights, free markets and free trade. Throughout the 1980s and 1990s, the Peruvian economist Hernando de Soto (2000), director of the neoliberal think tank *Instituto Libertad y Democracia*, had much influence on the economic policies in Peru. Accusing the state bureaucracy of being a barrier to development, he argued that creative individual entrepreneurs would have the chance to progress in a free market economy (de Soto 2000). For the small-scale farmers in Colca and Majes, however, the new 'freedoms' led instead to individualised risk, more uncertainty and increased precarity (Stensrud 2019a).

In the 1990s, neoliberal reforms took off when President Alberto Fujimori (1990–2000) implemented an extreme version of the structural adjustment advocated by the 'Washington Consensus' and the World Bank (Gonzales de Olarte 1993; Solfrini 2001). The so-called 'Fuji shock' consisted of radical austerity measures: higher interest rates and taxes, and a slashing of price subsidies and social spending. The shock was followed by neoliberal restructuring of the economy, deregulation of markets, massive privatisation, tax and tariff reform, and incentives for international investment (Klarén 2000; Crabtree 2002). In agrarian policy, the neoliberal restructuring and reduction of the state in the 1990s entailed a total reversal of the agrarian reform; the cancellation of all forms of subsidies and special credit to farmers, and the dismantling of the state apparatus that dealt with rural development (Mayer 2009). Influenced by de Soto's (2000) ideas of individual title deeds as key to development, the 1995 Land Law promoted the titling of individual land ownership, abolished the previous upper limits on personal landholding and allowed the state to sell land currently in public ownership. The new law opened the way to the capitalisation of agriculture by enabling titleholders to raise mortgages (Crabtree 2002: 142). Fujimori furthermore deregulated the food markets by cancelling price controls, allowing these to be set by market forces. He also closed down the institution that controlled imports of food and fertiliser (ENCI), and eliminated the Agrarian Bank as a source of subsidised credit to producers (Crabtree 2002: 141). Today, the small-scale farmers in Colca borrow money from private lending companies at high interest rates in order to buy seeds, fertilisers, herbicides and insecticides. The farmers are, however, very aware of the high economic risk involved in food production because of the volatility of the market, and they increasingly describe agriculture as a lottery where you can either win or lose (Stensrud 2019a). Because of the state's shedding of responsibilities, individual farmers are forced to take the brunt of the risks involved in farming, and they are strongly encouraged to become entrepreneurs (see also Marin, this volume).

The use of water, which was nationalised in the 1969 General Water Law that supplemented the agrarian reform (del Castillo 1994), was also re-regulated in the 1990s, thus exacerbating the uncertainties of farming and stoking the farmers' fear of a future water crisis. The functions of the state institutions in the operation, maintenance and administration of the irrigation infrastructure were handed over to the water user associations (*Juntas de Usuarios*), in which farmers are organised. To finance this work the *Juntas* could charge a tariff in order to have financial autonomy from the state. In Colca Valley, the farmers reacted with disbelief and horror when the water tariff was introduced in the 1990s: why should they pay for water that was given to them by the mountains? In their view, the water was a living and life-giving being and it did not belong to the state or the *Junta*, but to the sentient landscape in Colca. Only after the *Junta* had shown in practice that they would support the farmers, not only with materials, but also with technical and moral support, did they agree to pay the tariff (see also Stensrud 2016).

The Fujimori government fired large numbers of state functionaries in the water administration and made several attempts at privatising water, yet failed due to strong opposition from the irrigation organisations (Oré and Rap 2009; Oré et al. 2009: 52–3). The attempt to commodify water was part of a global trend that started with the 1992 Dublin Statement, in which the need to recognise water as an economic good was declared (Franco et al. 2013). In 2009, President Alan García (2006–11) succeeded in passing a new law on water resources (*Ley de Recursos Hídricos*), which was justified by the threat of climate change, population growth and increased mining activities (ANA 2010). Although water is still explicitly acknowledged as state property, the law has been adapted to a neoliberal project that induces the state to downplay its role in economic and social politics, creating opportunities for private companies to intervene and invest in hydraulic infrastructure (del Castillo 2011). Through individual licences for water-use rights and payments for the use of water, the law fosters individualised water management (Paerregaard et al. 2016). As the responsibility for sustainable management is being distributed among individual citizens, small-scale farmers become more vulnerable to changes in the environment and in the markets for agricultural products. Hence, vulnerability has more to do with social structure and economic politics than with natural processes or events (Marin, this volume), and it can be exacerbated by a water regime that favours some users and uses over others or heightens competition (Lynch 2012). As the state prioritises large-scale projects of irrigation and agribusiness, like the Majes Irrigation Project in the lower part of the Majes-Colca watershed, small-scale organic farming practices become virtually impossible to sustain.[3]

## 'MONEY CANNOT BE EATEN': CLIMATE CHANGE AND LOST HARVESTS

All over Andean Peru, farmers and scientists have reported changes in temperature, precipitation, seasonality, glacier retreat and water supply, which are seen as the result of global warming (Bates et al. 2008; Vuille et al. 2008; Bolin 2009). Peasant farmers, however, do not experience climate change as an isolated phenomenon, but as one of many issues that smallholders have to deal with. Several scholars have shown how changes in glaciers and water supply are entangled in socio-economic structures and neoliberal policies (Boelens 2009; Oré et al. 2009; Carey 2010; Lynch 2012; Rasmussen 2015). This is also the case in Colca Valley, where the rainy season of 2014 was ruined by drought and frost, and most farmers lost parts of their harvests of potatoes, beans, peas and maize. In Lari district, each of the 320 farmers had lost on average 60 per cent of their harvest. 'We are in a crisis all over the valley', the mayor of Lari said when I interviewed him in March 2014. He explained the crisis with the fact that most farmers had borrowed money from micro-finance agencies and now they would have to refinance their loans. Then, he elaborated on the climatic changes that they were experiencing:

> The cold, the snowfall, and also the excessive heat, increased temperatures that from 9 in the morning to 1 in the afternoon rise to almost 30 degrees [Celsius] ... The irrigated fields do not resist this tremendous heat, and the plants get stressed and wither, and in addition to this, at night the cold comes, the frost, and finishes killing the poor plant. Hence, production reaches only 50 per cent and this affects the economy of the people in this place, where 98 per cent are dedicated to agricultural work: cattle, sheep and the production of maize, potatoes, peas, beans and other vegetables that are vulnerable to cold weather. This is something unusual; we have never been presented [with] something that drastic. We have had cold spells, but they were benign, or mild, we could say. We could in one way or another reactivate the plant with irrigation, or with stimulations of vitamins and all, but at this time it is not possible, because the frost ... during three or four days, it has been devastating, definitive! ... The water [scarcity] can still be relieved, but the frost is devastating. It is an incredible burning. We have seen the fields. All is burnt ... [by the frost]. It is incredible, but it is really very serious.

In my first fieldwork in Colca Valley in 2011, farmers told me about their worries regarding the changing weather patterns: irregular incidents of frost, drying springs and pastures, and irregular rain – belated rain, less rain, torrential rainfall and rainfall in the dry season. Some farmers predicted a dark and dry future: that the water would disappear, and no life would be possible (Stensrud 2015). When I returned to Colca after two years absence in November 2013, the rainy season was supposed to start soon, but once again all the farmers were waiting in vain. 'The world is upside-down', one of my neighbours in Chivay, Miriam, said when she was telling me about the abnormal seasons in the past year: it had rained in the dry season, and the frost continued into (the supposed) rainy season. Like most farmers, she and her husband Pedro perceived the effects of climate change in terms of seasonal instability, belated rains, longer drought periods, disappearing mountain glaciers and decreasing water supply. The most notable changes were the sudden frosts that came when they were not supposed to, and the more extreme shifts in temperature: 'the sun seems to be closer to the earth', Miriam said. Pedro complained that they must irrigate more often because the heat made the earth dryer than usual. Miriam said: 'earlier the seasons were respected ... but the climate has changed and it affects the agriculture. The earth has transformed too much.' She was worried that their bean plants would get the 'chocolate illness' that blackens and ruins the beans. Miriam and Pedro had lived several years in Arequipa city, where the educational opportunities for their children were better. Now, they had moved back to Chivay to cultivate their fields and have a quiet and good life. However, they found that farming was not like it used to be when they were younger; the ecological and socio-economic conditions had changed. In order to survive economically, they had to find additional incomes. Miriam had a market stall where she sold clothes, some of which she embroidered herself in typical Colca patterns, and Pedro did some odd jobs for the municipality and others.

It is becoming increasingly difficult to earn a living as a small-scale peasant farmer, and many fields in Colca lie abandoned because 'it is not profitable' due to climatic uncertainties, falling product prices and increased costs of labour. Earlier, farmers in the valley usually performed tasks like sowing and harvesting collectively and reciprocally: farmers helped each other and mutually exchanged services. Today, it is more common to hire landless people – often migrants from highland communities – as day labourers (*peones*). Agriculture is still the main economic activity in Colca, but it has been undergoing a transition from subsistence farming to market production of potatoes, quinoa, beans and maize. 'We used to cultivate to consume, but now everything is money', Miriam said. Private banks and financial institutions offer micro-credit and loans to farmers at high interest rates of between 18 and 30 per cent. The money is spent on fertilisers, pesticides, herbicides and new seeds. New standardised types of potatoes have to a large degree replaced local varieties. In 2014, farmers in Chivay told me about a new potato called *La Única*, which had been introduced five years earlier. While they used to harvest potatoes seven months after sowing, *La Única* can be harvested after only four or five months in the highland climate, and therefore it is more efficient to produce for market sale. However, many farmers pointed out that *La Única* does not taste good and it cannot be preserved for a long time. The smaller potatoes they used to sow earlier, like the black potato, *Sika*, had a better flavour and could be kept for longer.

Buying new seeds and fertilisers is expensive, and farmers often take up credit loans before sowing. If the harvest fails, they are left with nothing but debt. After losing her harvest two years in a row, Hilda was paying off her debts by selling clothes in Chivay, where she rented a market kiosk beside Miriam's business. When I talked to her in November 2013, she was considering whether she should sow potatoes again. Potatoes are usually sown in the beginning of the rainy season, but the rain had not yet come. The fluctuating market price was another factor to take into consideration. Hilda said: 'When there is a lot of potatoes the price falls, and when there is little potatoes, the price increases. And the farmer cannot guess …' When I met her again in March, after four months of drought and frost periods, she told me that she had not sown anything this year and had instead chosen to focus on her clothes business in the Chivay market. It seemed like a good decision: her neighbours had lost a large part of their beans and all the potatoes were ruined. People were saying there would not be any potatoes this year.

As farmers make plans for the coming season, they have to anticipate the weather and the market. Both have become increasingly difficult to predict in the past couple of decades: the weather is more unstable because of climate change and market prices are more volatile because of deregulating policies. The ones that are making money on food are not the farmers, but the middlemen who buy directly from the farmer in the field and sell to wholesalers in urban and foreign markets. The leader of the main water-user association in Colca Valley (*la Junta de Usuarios Valle del Colca*) expressed his concern that all the young people wanted to be merchants and businessmen instead of producing food and that a large part of the fields had been abandoned. 'What are we going to eat?', he asked rhetorically; 'money cannot be eaten'. When people in Chivay must buy expensive potatoes imported from elsewhere, it means that life has changed beyond recognition and becomes unbearable for many.

## THE QUINOA BOOM, AGROCHEMICALS AND FOOD SAFETY

Many farmers in the valley have left their fields and moved to the city of Arequipa or to the Majes Irrigation Project in the lower part of Caylloma province, to find work and business opportunities. Those who stay are forced to change their farming practices by planting new seeds and using agrochemicals. In my 2013–14 fieldwork, Peruvian farmers were in the middle of the 'quinoa boom'. During the past decade, the quinoa seed has gained worldwide popularity as a high-protein and gluten-free food with extraordinary nutritional value, and the demand from US and European markets has skyrocketed. From 2010 to 2014, the global price doubled from US$2.96 to US$6.74 per kilogram (Statista 2018). Many farmers in Colca started to sow quinoa, hoping to get as much out of the high prices as they could. Another reason to sow quinoa is that the plant grows in difficult weather conditions and is known by farmers to resist both rain and frost; hence, it was also believed to be resistant to climatic change.

Quinoa is traditionally a highland plant that has been cultivated for thousands of years in the Andes, where it is grown organically with natural fertiliser (manure). The organic quinoa from the highlands is supposedly more valuable than the quinoa produced on the coast. However, based on their experience as small-scale peasant farmers, the farmers in Colca have little trust in the market system and the middlemen. The leader of the *Junta* in Colca put it bluntly: 'I don't trust this. They [private companies or middlemen] promise 20 *soles* [US$7] per kilogram, but when the moment comes, they only pay you 5 *soles*.' This is not profitable in the Colca Valley, where the yield is 1,500–2,000 kg of quinoa per hectare, and it takes seven months from when it is sown until it is ready to be harvested. Average landholding size is 1.2 hectares and many families own just a third of a hectare. However, a few farmers have scaled up production by renting land from others and applying artificial fertilisers and pesticides.

The Colca farmers also have to compete against medium- and large-scale farms and agribusinesses in irrigated areas on the Peruvian coast that started to cultivate quinoa as the global demand soared. In the Majes Irrigation Project, quinoa was sown on 2,000 hectares of land in 2013. Due to the warm climate, quinoa matures much faster and gives larger yields in Majes than in the highlands. The Inia Sansel variety matures in four months and the Sajama variety matures in only three months, and a farmer can get between 6,500 and 8,000 kg of quinoa per hectare. Farmers in Majes were paid 10 *soles* (US$3.5) per kilogram by the middlemen who came to their farms to buy quinoa in 2014.[4] Quinoa cannot be grown organically in Majes because of the increased incidence of plant disease and fungus (mildew) in the coastal climate. One farmer in Majes explained that they start by applying chemicals to the soil before sowing the quinoa. First, you have to drip irrigate for five or six days until the weeds start sprouting. Then you apply herbicide to kill the weeds, so that it will not mix with the quinoa. Only then it is ready to sow. The farmers themselves say that the quinoa becomes more 'synthetic' (*sintético*) because of all the chemicals.

During 2012–13, food safety appeared on the agenda in Peru, and also in Majes and Colca Valley, where people started to worry about the increased use of insecticides and artificial fertilisers. According to the farmers I interviewed in Colca, the chemicals destroy the soil and make it unproductive for three or four years. Some farmers have also started to use hormones, or 'vitamins' as they call them, in order to get bigger crops faster. The 'hormones' are mostly used on potatoes, to make them bigger and more attractive. However, most people are getting worried about this trend. One man said: 'It is not natural anymore. They put everything in [the plants]: insecticide, fertilizers, hormones, and against the frost they use something that makes the plant rise up again. That is not natural. That is junk food [*comida chatarra*]!' The people I talked to in Colca tended to attribute new illnesses – like cancer – to the chemicals and hormones used in food production, which created a lot of uncertainty and fear. The mayor of Lari village, who was also a farmer, told me that 'the [agrochemical] industries come here to our district, promoting their products and saying that everything is good … and we don't have evidence, but it has contaminated us'. I was frequently told that chemicals and hormones were supposed to help plants survive frost. According to the Lari mayor, however, plants that have been given natural manure have resisted the frost better than plants given artificial fertilisers: 'the agrochemicals nurture in a fanciful way … it hurts the plant, which has no resistance against the cold'.

Many of the people I met in Colca, like Miriam and her friends and marketplace colleagues Cirila and Hilda, tended to look to the past to find healthier alternatives. Cirila said:

> Earlier they used natural fertilizers from the cattle pen, but now they use [artificial] fertilizers and hormones. That's why the potatoes get big. And this produces cancer. The cancer is due to the fertilizers, the hormones they put in. Those potatoes are not like before, and they don't have the same flavour like the small potatoes that we used to eat.

Miriam remembered her childhood: 'It used to be calmer … The nutrition was healthier. They did not use insecticides; they used ashes from the *yareta*, which they used as firewood.⁵ Now there is almost no *yareta* left. It grows like green mushroom in the hills and it is good to burn.' Hilda added that the *yareta* 'grows like a human being; it takes time to grow'. Earlier generations seemed to have time to grow small potatoes and there was enough slow-growing *yareta* for everyone. Today, farmers must not only adapt to a changing environment, but they must also adjust to constantly changing markets with volatile prices and they have to look for additional incomes. This requires efficiency and flexibility in a world that is changing faster than it used to (Eriksen 2016), and many people lament these recent changes. 'The micro-finance agencies [*micro-financieras*] have made us forget certain customs', the mayor of Lari said. He explained that in his opinion, farmers should go back to doing communal work and *ayni*, in order to stand united as equals: 'this is a way of mitigating'. He was not only referring to the mitigation of climate change, but to the effects of market liberalisation and price volatility, which creates winners and losers among the farmers. A proposal to return to *ayni* – which is a form of labour exchange based on mutual reciprocity (Allen 2002; Mayer 2002) – is a way of envisioning the future as in need of commonality and mutuality. The mayor's opinion resonates with the Quechua concept of *sumaq kawsay*, translated to *buen vivir*, or 'good life', which conveys the idea of living in plenitude, in a state of respect and balance between the individual, society and cycles of nature (Merino 2016: 276). Although *buen vivir* is not incorporated into Peruvian legislation, as it has been in Ecuador and Bolivia, it is emerging as an alternative to the mainstream development paradigm and a platform for political articulations (Merino 2016). In private spaces, I often heard Colca farmers criticise the hectic urban lifestyle in the cities, where the constant pursuit of money has made people stressed and forgetful of the importance of maintaining relationships with kin and the landscape. While some farmers in Colca look to the past in order to find ways to adapt to their changing environment, the government continues to insist on the importance of progress and economic growth. State programmes tend to encourage modernisation, seen as commercialisation, supporting training in marketing, entrepreneurship and the use of technology, as strategies to overcome poverty in the highlands and to create progress.

## MODERNISATION OF TECHNOLOGY AS CLIMATE CHANGE ADAPTATION

The government prioritises investments in large-scale export-oriented agribusiness on the coast, as well as mining and other extractive industries, in the name of progress and economic growth. The idea of progress has become a hegemonic narrative in Peru, where it guides political priorities and public investments, yet the meaning of progress is contested and the benefits of Peru's economic growth are unevenly distributed. One of my interlocutors, an agricultural engineer born and raised in Chivay, the main town in Colca Valley, expressed his concern: 'There is no growth in Chivay. In Peru there is economic growth, but where is that growth? It has not arrived here. Maybe it is in construction and housing, or in mining, but not in agriculture.' Chivay and the other villages in Colca Valley experience a declining agriculture and out-migration of young people. According to the dominant narrative in Peru, the Andes is lagging 'behind' the modernised coast in the imaginary of linear development: people there are seen as needing time, investment and education to catch up with the developed urban coast.

Today, the farmers and pastoralists in the headwaters are the first to experience the consequences of climate change and the last to benefit from economic growth. They are the first to lose out in the competition with larger producers, and they are also the first to be categorised as victims because of their supposed 'backwardness'. As victims, they are defined as recipients of development aid with the aim of making them more 'modern', but they are not allowed to participate in the design of the projects and programmes that are supposed to help them. In spite of the talk of participation, dialogue and mutual exchange, the development and adaptation programmes are designed by technocrats in state institutions and non-governmental organisations (NGOs) with the explicit purpose of modernising farming: in other words, farmers should become more entrepreneurial, use newer technology and be more integrated in the market economy. However, as early as 1944, Karl Polanyi warned against the danger of relying on the free market: 'to allow the market mechanism to be sole director of the fate of human beings and their natural environment indeed, even of the amount and use of purchasing power, would result in the demolition of society' (Polanyi 1944 [2001]: 76).

In the past two decades, narratives about global climate change have become more widespread in Colca Valley as they are disseminated by governmental and NGOs in talks and workshops. Government institutions and international development agencies increasingly finance projects related to climate change adaptation in the Andes. However, these institutions and agencies tend to reduce the complexity of climatic and economic change to manageable issues such as lack of modern technology. One of the recent development programmes in Colca Valley is the *Programa Subsectorial de Irrigaciones Sierra* (PSI): a programme for the modernisation of irrigation technologies in the Peruvian highlands.

Funded by the World Bank, the programme started in Colca Valley in 2011. Colca Valley was chosen because the majority of the farmers there pay the water tariff, and this was seen as evidence of responsibility. By encouraging farmers to transition from traditional surface furrow irrigation to modern engine-driven sprinkler and drip systems, the goal was to achieve higher productivity and profitability in agriculture through modern technology and efficient water use. The PSI engineers used the threat of global warming and permanent water scarcity to argue that modern technology is the best way to adapt to climate change. In this way, the complexity of the changing environment, the free market economy, poverty and migration were reduced to issues of water availability, and technology was presented as an easy fix (cf. Li 2007). The technology is expected to introduce efficiency and market-oriented production in the rural highlands. In line with the neoliberal ideology that was consolidated in Peru in the 1990s, to be 'modern' today also means to be entrepreneurial and business-oriented. Development programmes in Peru often aim to complete the marketisation of 'non-modern' practices, and PSI is yet another example of a global trend where environmental regimes are based on the capitalist idea of continued economic growth (Wilhite and Salinas, this volume). I suggest that PSI will push farmers in Colca Valley further into market relations and dependency on monetary income and the companies that provide the technology (see also Stensrud 2019b).

## CONCLUSION: THE DARK SIDE OF 'PROGRESS'

The storm irresistibly propels him [the angel of history] into the future to which his back is turned, while the pile of debris before him grows skyward. This storm is what we call progress. (Benjamin 1968: 257–8)

The accelerating winds of modernisation and what we call 'progress' blow in all directions, creating wealth for a few and poverty and destruction for many others. As in the rest of the world, economic growth is the main goal for the Peruvian government, and parts of 'nature' – water, land, forest and minerals – are seen simply as economic resources and a means to create growth. Although the growth is based on extractivism, the hegemonic narrative of progress and development is not associated with debris in Peru; on the contrary, the desire for progress is undisputable and critique is almost impossible. Those who criticise this modernity tend to be deemed ignorant and provincial.

Too often, complex ecological changes are reduced to singular problems that can be readily defined and diagnosed in development reports and thus seen as easily managed and fixed by techno-optimistic solutions (Haraway et al. 2016). As Obertreis et al. (2016: 170) argue, 'the main thrust in the neoliberal era is the use of market mechanisms and technological fixes as a solution to environmental problems'. In the PSI programme in Colca Valley, the discourse of climate change is used to impose technological solutions and modernisation. The Peruvian state's modernisation project is not neutral, however, but embedded in socio-economic and racial hierarchies. As many scholars have pointed out, modernity in Latin America cannot be understood as separate from coloniality. The modern world is built on colonialism, and colonial structures of power – based on ideas of race and control over labour and resources – are still constitutive for today's modernity (Mignolo 2000; Quijano 2000; Escobar 2007). Mignolo (2000: 22) understands 'coloniality' as 'the reverse and unavoidable side of "modernity" – its darker side, like the part of the moon we do not see when we observe it from earth'. Hence, I suggest that global climate change and the slow violence of ecological crisis cannot be understood as separate from coloniality, especially in Latin America. The people who are living in the climate-sensitive areas in the Peruvian highlands have been suffering from economic inequality and discrimination for centuries, and today they suffer the consequences of climate change. They are the principal casualties of slow violence: the disasters that are relatively invisible and of little interest to the sensation-driven media (Nixon 2011). Paradoxically, this slow violence is partly caused by the accelerating speed of economic globalisation.

I have in this chapter argued that the effects of climate change as well as the effects of capitalisation and deregulation of agriculture are making it harder for small-scale farmers in Peru to maintain their livelihood. Changes in the environment and the economy mutually exacerbate each other and reinforce the tendency of abandonment of the rural highlands and migration to the coastal cities. For example, the use of agrochemicals in Colca Valley started as a result of market liberalisation and increased competition, but it is also used to combat new plant diseases resulting from warmer temperatures. Today, people are worried about the effects that these chemicals might have for the soil, food and health. In addition, a consistent lack of long-term planning and preventive measures as well as a dismantling of collective solutions have worsened the consequences of climate change. Farmers are forced to take risks by guessing which crops to sow and taking up loans. In other words, the neoliberal state sheds responsibility by making farmers shoulder the risks as individual entrepreneurs. Poverty and insecure access to water are in the public discourse often defined as problems related to climate change and a lack of 'culture' and education, and are thus not seen as related to socio-economic structures, coloniality and the reduction of public welfare. The modern project has from the start separated nature from culture and insists on perceiving nature as a repository of resources to use for the extraction of monetary value (Tsing 2015; West 2016). The hegemonic project of modernisation and progress as it is implemented by the recent and current Peruvian governments is akin to imitating the rich industrialised countries in the global North: investing in modern technology, promoting the individual, dismantling the state and liberalising the economy. These measures are presented as necessary for a better future in spite of the evidence of ecological destruction and the oppression of those who are not willing to adopt a modern way of life or who oppose an extractivist economy and the necessity of continued economic growth (Ødegaard and Rivera Andía 2019).

Walter Benjamin's 'angel of history' is forced into the future, to which his back is turned, while facing the past and watching the catastrophes and ruins caused by progress that keep piling up in front of him (Benjamin 1968). In the Quechua language, 'future' is called *qhepa*, which can be translated as both 'behind' and 'future', while *ñawpa* can be translated as both 'ahead' and 'past'. In other words, for us, as for the farmers of Colca Valley, the past is in front of us and the future is behind our backs. Looking at our past while we walk backwards into the future forces us to see the consequences of our industrialised, fossil fuel-dependent capitalist society. The ecological destruction is slow and incremental, but when it reaches the tipping point, the result is devastating. To counter this looming catastrophe, we need a better understanding of the complex interplay of climate, environmental and agricultural policies and the effects of the free market on land, water and people.

## NOTES

1. In 2014, 1 Peruvian *Nuevo Sol* was equivalent to US$0.35.
2. According to the National Population Census, Caylloma province had a population of 73,718 in 2007, and 86,771 in 2017. Of these, 26,663 lived in the highland districts, while the registered population in the lowland Majes district, comprising the Majes Irrigation Project, was 60,108 persons (INEI 2018).
3. The Majes Irrigation Project takes water from the Condoroma Dam in the highlands, into the Majes Canal (through Colca Valley) and down to the arid pampa of Majes. It is managed by a public agency, and the government spends annually US$33 million on infrastructure insurance and staff (see Vera Delgado and Zwarteveen 2008).
4. In 2011, Peru harvested a little less than 35,475 hectares of quinoa, and in 2015, the harvested area of quinoa had grown to 69,303 hectares (Statista

2019), making Peru the world's major exporter. After this increment in production, the prices started to fall, and in 2015 the price of quinoa fell to 3.5 soles (approximately US$1) per kilogram, compared to 11 soles in 2013 and 10 soles in 2014 (Hudson 2015).

5. *Yareta* (*Azorella compacta*) is an evergreen perennial plant in the Apiaceae family that grows in the Andean highlands. Since it has traditionally been used as fuel and because of its slow growth, it is in danger of extinction in many places.

## REFERENCES

Allen, C.J. 2002. *The Hold Life Has. Coca and Cultural Identity in an Andean Community*, second edition. Washington, DC: Smithsonian Institution Press.

ANA (Autoridad Nacional del Agua). 2010. *Ley De Recursos Hídricos y su Reglamento*. Ley No. 29338. LIMA: Ministerio de Agricultura.

Bates, B.C., Z.W. Kundzewicz, S. Wu and J.P. Palutikof (eds). 2008 *Climate Change and Water*. Geneva: IPCC Secretariat.

Benjamin, W. 1968. *Illuminations*, ed. and introduction H. Arendt. New York: Schocken Books.

Boelens, R. 2009. The Politics of Disciplining Water Rights. *Development and Change* 40(2): 307–31.

Bolin, I. 2009. The Glaciers of the Andes are Melting: Indigenous and Anthropological Knowledge Merge in Restoring Water Resources. In S.A. Crate and M. Nuttall (eds), *Anthropology and Climate Change: From Encounters to Actions*. Walnut Creek, CA: Left Coast Press, pp. 228–39.

Buho, El. 2013. Provincia de Caylloma es Declarada en Emergencia, 2 March. Available at: https://elbuho.pe/2013/03/provincia-de-caylloma-es-declarada-en-emergencia/ (accessed 10 January 2019).

Carey, M. 2010. *In the Shadow of Melting Glaciers: Climate Change and Andean Society*. New York: Oxford University Press.

Collier, D. 1978. *Barriadas y Élites: de Odría a Velasco*. Lima: IEP.

Crabtree, J. 2002. The Impact of Neo-liberal Economics on Peruvian Peasant Agriculture in the 1990s. *The Journal of Peasant Studies* 29(3–4): 131–61.

de la Cadena, M. 2000. *Indigenous Mestizos: The Politics of Race and Culture in Cuzco, 1919–1991*. Durham, NC: Duke University Press.

de Soto, H. 2000. *The Mystery of Capital: Why Capitalism Triumphs in the West and Fails Everywhere Else*. New York: Basic Books.

del Castillo, L. 1994. Lo Bueno, lo Malo y lo Feo de la Legislación de Aguas. *Debate Agrario* 18: 1–20.

—— 2011. Ley de Recursos Hídricos: Necesaria pero No Suficiente. *Debate Agrario* 45: 91–118.

Eriksen, T.H. 2016. *Overheating*. London: Pluto Press.

Escobar, A. 2007. Worlds and Knowledges Otherwise. *Cultural Studies* 21(2–3): 179–210.

Ferguson, J. 1994. *The Anti-politics Machine. 'Development', Depoliticization, and Bureaucratic Power in Lesotho*. Minneapolis, MN: University of Minnesota Press.

Franco, J., L. Mehta and G.J. Veldwisch. 2013. The Global Politics of Water Grabbing. *Third World Quarterly* 34(9): 1651–75.

Gonzalez Casanova, P. 1965. Internal Colonialism and National Development. *Studies in Comparative International Development* 1(4): 27–37.

Gonzales de Olarte, E. 1993. Economic Stabilization and Structural Adjustment Under Fujimori. *Journal of Interamerican Studies and World Affairs* 35(2): 51–80.

Haraway, D., N. Ishikawa, S.F. Gilbert, K. Olwig, A.L. Tsing and N. Bubandt. 2016. Anthropologists are Talking – About the Anthropocene. *Ethnos* 81(3): 535–64.

Harvey, D. 2005. *A Brief History of Neoliberalism*. Oxford: Oxford University Press.

—— 2006. *Spaces of Global Capitalism: Towards a Theory of Uneven Geographical Development*. London: Verso.

Hastrup, K. 2009. Arctic Hunters: Climate Variability and Social Flexibility. In K. Hastrup (ed.), *The Question of Resilience: Social Responses to Climate Change*. Copenhagen: Royal Danish Academy of Sciences and Letters, pp. 245–70.

Hudson, L. 2015. Quinoa Prices Fall (Finally) Due to Rise in Production. SpendMatters, 27 April. Available at: http://spendmatters.com/2015/04/27/quinoa-prices-fall-finally-due-to-rise-in-production/ (accessed 18 February 2019).

INEI. 2018. *Resultados Definitivos de los Censos Nacionales 2017, Arequipa, XII de Poblacion, VII de Vivienda y III de Comunidades Indígenas*. Lima: Instituto Nacional de Estadística e Informática.

Klarén, P.F. 2000. *Peru: Society and Nationhood in the Andes*. New York: Oxford University Press.

Li, T.M. 2007. *The Will to Improve. Governmentality, Development, and the Practice of Politics*. Durham, NC: Duke University Press.

Lynch, B.D. 2012. Vulnerabilities, Competition, and Rights in a Context of Climate Change Toward Equitable Water Governance in Peru's Rio Santa Valley. *Global Environmental Change* 22: 364–73.

Massey, D. 2006. Space, Time and Political Responsibility in the Midst of Global Inequality. *Erdkunde* 60(2): 89–95.

Mayer, E. 2002. *The Articulated Peasant: Household Economies in the Andes*. Boulder, CO: Westview Press.

—— 2009. *Ugly Stories of the Peruvian Agrarian Reform*. Durham, NC: Duke University Press.

Mignolo, W.D. 2000. *Local Histories/Global Designs: Coloniality, Subaltern Knowledges, and Border Thinking*. Princeton, NJ: Princeton University Press.

Merino, R. 2016. An Alternative to 'Alternative Development'?: Buen Vivir and Human Development in Andean Countries, *Oxford Development Studies* 44(3): 271–86.

Nixon, R. 2011. *Slow Violence and the Environmentalism of the Poor*. Cambridge, MA: Harvard University Press.
Obertreis, J., T. Moss, P.P. Mollinga and C. Bichsel. 2016. Water, Infrastructure and Political Rule: Introduction to the Special Issue. *Water Alternatives* 9(2): 168–81.
Ødegaard, C.V. and J.J. Rivera Andía (eds). 2019. *Indigenous Life Projects and Extractivism: Ethnographies from South America*. Cham, Switzerland: Palgrave Macmillan.
Oré, M.T. and E. Rap. 2009. Políticas Neoliberales de Agua en el Perú. Antecedentes y Entretelones de la Ley de Recursos Hídricos. *Debates en Sociología* 34: 32–66.
Oré, M.T., L. del Castillo, S. Van Orsel and J. Vos. 2009. *El Agua, ante Nuevos Desafíos: Actores e Iniciativas en Ecuador, Perú y Bolivia, Agua y Sociedad*. Lima: Instituto de Estudios Peruanos.
Paerregaard, K., A.B. Stensrud and A.O. Andersen. 2016. Water Citizenship: Negotiating Water Rights and Contesting Water Culture in the Peruvian Andes. *Latin American Research Review* 51(1): 198–217.
Peru.com. 2011. Declaran en Emergencia Provincia de Caylloma, 17 March. Available at: from: https://peru.com/2011/03/17/actualidad/otras-noticias/declaran-emergencia-provincia-caylloma-noticia-692 (accessed 10 January 2019).
Polanyi, K. 2001. *The Great Transformation: The Political and Economic Origins of Our Time*. Boston, MA: Beacon Press. Originally published 1944.
Quijano, A. 2000. Coloniality of Power, Eurocentrism, and Latin America. *Nepantla: Views from South* 1(3): 533–80.
Rasmussen, M.B. 2015. *Andean Waterways. Resource Politics in Highland Peru*. Seattle, WA and London: University of Washington Press.
RPP Noticias. 2012. Arequipa: Declaran en Emergencia la Provincia de Caylloma por Lluvias, 13 February. Available at: https://rpp.pe/peru/actualidad/arequipa-declaran-en-emergencia-la-provincia-de-caylloma-por-lluvias-noticia-450725 (accessed 10 January 2019).
—— 2013. Arequipa: Declaran en Emergencia Provincia de Caylloma, 1 March. Available at: https://rpp.pe/peru/actualidad/arequipa-declaran-en-emergencia-provincia-de-caylloma-noticia-572158 (accessed 10 January 2019).
—— 2014. Arequipa: Sequía y Heladas Dejan Pérdidas por 14 Millones en Caylloma, 18 February. Available at: http://rpp.pe/peru/actualidad/arequipa-sequia-y-heladas-dejan-perdidas-por-14-millones-en-caylloma-noticia-670750 (accessed 26 April 2017).
Seligmann, L.J. 1995. *Between Reform and Revolution. Political Struggles in the Peruvian Andes, 1969–1991*. Stanford, CA: Stanford University Press.
Smith, N. 1984. *Uneven Development: Nature, Capital, and the Production of Space*. Oxford: Blackwell.
Solfrini, G. 2001. The Peruvian Labor Movement Under Authoritarian Neoliberalism. From Decline to Demise. *International Journal of Political Economy* 31(2): 44–77.

Statista. 2018. Quinoa Market: Statistics & Facts. Available at: www.statista.com/topics/2813/quinoa-market (accessed 19 November 2018).

—— 2019. Harvested Area of Quinoa in Peru from Crop Year 2010 to 2017 (in Hectares). Available at: www.statista.com/statistics/518497/quinoa-harvested-area-peru/ (accessed 18 February 2019).

Stensrud, A.B. 2015. Raining in the Andes: Disrupted Seasonal and Hydrological Cycles. In K. Hastrup and F. Hastrup (eds), *Waterworlds: Anthropology in Fluid Environments*. London and New York: Berghahn Books, pp. 75–92.

—— 2016. Climate Change, Water Practices and Relational Worlds in the Andes. *Ethnos: Journal of Anthropology* 81(1): 75–98.

—— 2019a. Safe Milk and Risky Quinoa: The Lottery and Precarity of Farming in Peru. *FOCAAL* 83: 72–84.

—— 2019b. 'You Cannot Contradict the Engineer': Disencounters of Modern Technology, Climate Change and Power in the Peruvian Andes. *Critique of Anthropology*. doi: 10.1177/0308275X18821164.

Tsing, A. 2015. *The Mushroom at the End of the World: On the Possibility of Life in Capitalist Ruins*. Princeton, N.J.: Princeton University Press.

Vera Delgado, J. and M. Zwarteveen. 2008. Modernity, Exclusion and Resistance: Water and Indigenous Struggles in Peru. *Development* 51: 114–20.

Vuille, M., B. Francou, P. Wagnon, I. Juen, G. Kaser, B.G. Mark and R.S. Bradley. 2008. Climate Change and Tropical Andean Glaciers: Past, Present, and Future. *Earth Science Reviews* 89: 79–96.

West, P. 2016. *Dispossession and the Environment: Rhetoric and Inequality in Papua New Guinea*. New York: Columbia University Press.

# 7. Puzzling Pieces and Situated Urgencies of Climate Change and Globalisation in the High Arctic
## Three Stories from Qaanaaq

*Astrid Oberborbeck Andersen and Janne Flora*

### INTRODUCTION

This chapter traces situations in which economic and material practices that can be connected to a globalised economy intersect with elements connected to climate change. We present three stories from the High Arctic, each showing moments where such intersections happen in different ways, generating specific concerns and lines of effects. The stories are situated in Qaanaaq, the northernmost town in Greenland, located at 77° N. Here, although human livelihoods and ecosystems are affected by globalised economies – capitalist modes of production and lines of consumption – people live with limited access to free flows of money and commodities. Limited in the sense that the spending capacity of people in Qaanaaq and neighbouring settlements is not the only factor determining how much can be consumed and how much circulates. Commercial goods from and to the outside world arrive and leave in two annual supply ships, in July and September. In addition, an oil tanker brings supplies once a year. To a smaller extent, one weekly airplane and a few helicopters transport people, mail and small amounts of foodstuffs or other goods to Qaanaaq. The majority of these supplies is stored and sold in the state-owned *Pilersuisoq*, the one all-purpose store in Qaanaaq. Long-life milk, butter and a small assortment of other dairy products, mainly Danish brands, are sold here, as are potatoes, onions, carrots and a few other fresh vegetables that arrive with the ships. Other vegetables, fruits, meats and fish are all frozen. The shop also sells tobacco, coffee and teas, canned and bottled drinks, and candy, household appliances, toys, clothing and toiletries and supplies for fishing and hunting, and dog food. The store stays well equipped for some months after supplies have arrived, but by Christmas time, some crucial ingredients might be lacking. The following months, the variety on offer steadily reduces –

until the next ship arrives. *Pilersuisoq*, as well as the company that ships the goods, Royal Arctic Line, are owned by the Greenland government, and thus the state participates in the circulation of most goods. Access to commercial goods is marked by seasonality, as are the hunting and fishing grounds that sustain many families in Qaanaaq.

Global warming has resulted in increasingly shorter seasons for fast sea ice during which significant parts of subsistence hunting and commercial fishing take place. Approximately 650 people live in Qaanaaq (Hastrup 2013a; Nissen et al. 2017); in 2015, there were 63 registered full-time hunters, a number that has been decreasing steadily since the early 1990s. Middle-aged people remember a season with fast sea ice stretching from September until July, providing a solid infrastructure for travel by dog sledge and hunting. Nowadays, the sea ice settles in November or December, and may break up several times due to storms, before it is thick enough to support dog sledge travel and fishing from the ice. While changes in the environment unsettle the predictability and seasonality of hunting practices, new horizons for monetary income are appearing, such as fishing and tourism (Flora et al. 2018). Yet, the flow of goods and money, and also the opportunities for producing dividends, are limited due to the high costs of transportation to and out of this region. These conditions make Qaanaaq an interesting site to study human-environment-capital relations in a world that is undergoing accelerated change (Eriksen 2016).

This chapter aims to describe and situate particular concerns and effects related to climate change, and to analyse how climate change effects tangle with and impinge upon the complexity of local livelihoods and national aspirations in Greenland. It questions how public and political discourses tend to engage with climatic and environmental change as *one* phenomenon that takes place globally, and poses *a* puzzle to be solved by a global community. Rather than seeing climate change as providing humanity with one broad puzzle to be solved, we propose these ethnographic stories as pieces of a puzzle, or puzzling pieces, each providing viewpoints and specifically situated challenges and concerns to learn from in order to deal with climatic changes and changes in livelihood opportunities, theoretically as well as practically and politically. In Greenland, as elsewhere, we suggest, different people take up different socio-economic positions and cultural viewpoints, and therefore they have specific priorities and strategies in how to deal with a changing climate and a changing economy.

Although the pieces are narrated from Qaanaaq, the materials present in them have different trajectories, in geographical and use terms. Story 1 explores economic practices by following crafted goods, food, new and used clothing as they are exchanged locally and along the Greenlandic west coast. Story 2 explores how Qaanaaq sits within a national and global economy by following the flow – and restriction on flows – of living resources. Story 3 demonstrates how climate change as a global concern makes Qaanaaq an iconic place, attracting people who arrive to document the melting ice and the last hunters living there. This generates a circulation of capital, a tourism sector, and the commodification of traditional livelihoods.

Through these ethnographic stories, the chapter moves between scales and engages with different perspectives by following economic practices and the circulation of materials. We show how the concerns and effects of climate change intersect with economic globalisation in myriad and complex ways. Second, we argue that when it comes to societal changes related to environmental and economic changes, actors positioned differently (geographically and socio-economically) emphasise differing priorities in terms of what is considered urgent. These differing priorities and urgencies demand different responses; how people act in the face of challenges varies significantly in relation to how actors are positioned, which in turn is linked to different livelihoods.

Adaptation, as well as climate change and capitalism, easily become general and singular concepts, in the sense of implicitly gathering many different kinds of processes and events under a common name, thus obscuring specificities. When speaking of climate change or globalisation in the singular, as general and universal phenomena, these acquire the quality of 'monster concepts'. Adaptation, just like vulnerability, mitigation and resilience, is part of the standard vocabulary used in climate change science literature and policies to address and respond to climate change (for example, Adger et al. 2005). We set out with a concern about how to engage with such monster concepts and phenomena as climate change and globalisation. This concern is related to translation and scale; what happens when we connect situations, materials and trajectories that take place in specific sites to analytical narratives that are generated in other contexts?

We suggest that one should be cautious when scaling ethnographic stories and in this way disconnecting them from the people and concerns where they originate. The danger in doing so lies in the potential violence done to local lives and livelihoods when decisions are made far away to solve or respond to problems of a general character. Rather than scaling up the effects of and responses to climate change, seeking to generate a whole picture (or approach) that can be used across contexts, we provide puzzling pieces, insisting that effects and responses to climatic changes and the global economy connect partially (Strathern 1991 [2004]). The vocabulary of partiality, and the resistance to generalise or scale phenomena is inspired by Marilyn Strathern, who in her book *Partial Connections* (1991 [2004]) argues that there is no such thing as parts and wholes, and that no perspective available to the analyst offers a totalising vista (1991 [2004]: xvi). One can switch perspectives, and also scale, but any switch of scale is problematic in that it creates a multiplier effect as well as a 'loss' of information (1991 [2004]: xv). The following stories are thought of as such puzzling pieces, aiming to show how the changing realities in Qaanaaq emerge distinctly in different constellations of practices, narratives and policies; partially connecting to realities of climate change, and partially to those of a globalised economy.

## STORY 1: TRACING TRADE AND MODALITIES OF CONSUMPTION. QAANAAQ–NUUK–EBAY CHINA–QAANAAQ

Jensine[1] was going home to Qaanaaq after some time in Nuuk, Greenland's capital, where she had attended professional training. We[2] met Jensine in the airport of Kangerlussuaq, an old US air base where most international flights enter and leave Greenland. We were travelling to Qaanaaq for fieldwork, and met Jensine unexpectedly in the waiting area where she was sitting with several bags of carry-on luggage. She pointed to one of them, full of things she had bought in the capital; presents for her six-year-old daughter's combined Halloween and birthday party the following week. These things could not be bought in Qaanaaq, and Jensine was happy to be bringing home special goods. The Halloween goodies had replaced other items in Jensine's luggage. When travelling southbound two weeks earlier, she had been carrying goods crafted by her parents: six dog sledge whips made by her father, a respected and elderly big hunter, two pairs of seal mitts decorated with a rim of polar bear fur and four bead necklaces crafted by her mother. 'All Thule style', Jensine smiled proudly, referring to the bend in the mitts, suited for travelling with dog sledge, and characteristic of the region in Northwest Greenland known as *Avanersuaq*, the 'big north'. Before travelling south, these crafted goods were advertised on Facebook, and when leaving her home, all items had already been sold to persons in the towns she would be passing through during the numerous flights towards Nuuk, approximately 1,600 km further south.

Jensine and her husband, both employed full time in the public sector and with stable incomes, cannot access such a variety of goods without travelling. Buying things on the Internet is mostly not an option, since the shipping costs are extremely high, unless they choose to receive their goods with the supply ship, which often means waiting for months. Jensine had, however, found a loophole; she told us that 'luckily, eBay in

China ships [in this case by air] products to the entire world for free'. The wallpaper covering one of the walls in Jensine's living room had made it to Qaanaaq from China, via eBay China, for an extremely low price.

If the *culture of consumption* is to be understood as 'an integral component of global capitalism' (Bodley in Baer 2012: 40), then this example of how Jensine makes different merchandise move in and out of Qaanaaq shows that 'the culture of consumption' indeed has many variations, depending on where in the world a person is positioned, what kind of consumer good is being put into circulation, and in what conditions it has been produced. Although commercialised goods have been shipped to and from Avanersuaq (also known as Thule) since the early 1900s (Hastrup, this volume), the people inhabiting the area are positioned on the edge of global flows of consumer goods, in what is often characterised as an economy based heavily on subsistence, circularity, sharing and subsidies (Nuttall 2000).

Once goods have made it from the outside to Qaanaaq, they, or their components, can have a long life, circulating between people, uses, homes and places, before ending their days in the junkyard or somewhere else. And even that might not be the end of things. The local store and Facebook both have a bulletin board where all kinds of items are put up for sale. Clothes, televisions, computers, mobile phones, rifles and fishing gear, as well as meat from hunting, are among the things frequently sold via these sites. Also, narwhal or walrus tusks, or locally crafted jewellery, tools and accessories made from these materials are sold on Facebook. Whereas the bulletin board in the store mainly reaches local buyers, Facebook posts reach a wider range of customers. In summer months, when narwhal is caught in the Qaanaaq district, before reaching hunting grounds further south, and before the supply ship has picked up the *mattak* (narwhal skin and blubber) bought by the local fish factory, it is sold for a high price via private networks, through Facebook or cell phones, to people along the west coast, and flown out in the weekly airplane.

Another site where goods land to later find new use is the dump, a little to the west of the town. This is where Qaanaaq's trash and obsolete items are deposited. The dump contains goods and materials: old washing machines, hospital furniture, dog sledges, cars, building materials, outboard motors and so on. People from Qaanaaq occasionally visit the dump to recover materials for particular uses; hunters may search the place for iron pipes that can be used to make new tools.

From Story 1 we learn that trade and consumption takes many forms. This insight is not new; there is a long history of economic anthropology discussing how local or indigenous forms of exchange and consumption have been impacted through colonisation and introduction of monetary systems of value and exchange (Polanyi 1957; Bohannan 1959; Taussig 1977), and how imperial or foreign modes of economy become embraced and embedded in local cultural values, moralities and power relations (Kopytoff 1986; Parry and Bloch 1989). Likewise, anthropological literature is rich in analyses of exchange relations, reciprocity and transformations of the value of objects and commodification (Gudeman 2001; Tsing 2013). Rather than entering into debate with this vast body of literature, I build on it to make a contribution to the literature on the consequences of climatic and environmental changes for local livelihoods. The argument in this regard is that climate change effects are never experienced in a vacuum, isolated from economic and political changes.

Social science literature about economic life in Inuit Arctic is mainly concerned with the way the traditional subsistence and sharing economy, based on hunting and fishing, is combined with a cash economy in specific communities (for example, Wenzel 2000; Collings 2010; Flora et al. 2018), and how kinship serves as an organising principle for activities of subsistence and exchange (Nuttall 2000; Flora 2019). Numerous accounts show that the introduction of a cash economy has produced increased inequality in communities in the Arctic (Harder and Wenzel 2012; Hastrup et al. 2018). Historical and archaeological accounts have emphasised that networks of exchange of goods between what otherwise seem to have been isolated groups of peoples in the Arctic have spanned vast areas for several centuries (Grønnow and Sørensen 2006; Grønnow 2016; Hastrup et al. 2018; Nuttall, this volume). Iron from meteorites, for instance, was hammered off with stone near Savissivik, south of Qaanaaq, and used in tools long before encounters with Europeans. Such pieces of meteoric iron have been found 1,000 km from their source, in the Western Arctic (Appelt 2004).

In present-day Qaanaaq, local consumption and trade is connected to global markets through eBay China and container ships, and to a national market through Facebook, social relations and cell phones. Yet, this connection is fragile due to the high cost of communication and transportation, unpredictable weather conditions and climatic variability. Further, consumption, and the way things change value as they move in and out of categories as merchandise, user goods, commodities, catch, meat and waste can be characterised as circular. This movement requires relations and skills that enable items and things to be acquired, used, transformed, deposited, distributed, re-crafted and used again.

Mark Nuttall, who has carried out research in Qaanaaq and communities further south, has richly described the ways in which kinship and relatedness inform social and economic life (Nuttall 2000). Although 'kinship as the organising principle for seal hunting, whaling, fishing, sharing and exchange is beginning to look precarious as Greenland moves increasingly towards a market-based economy' (Nuttall 2000: 35), Nuttall still situates kinship at the core of social and economic life in Northwest Greenland. Particular relations come with obligations of sharing and sets of behaviour, and how transformation of catch to meat to commodity to something else happens is not independent of kinship and other social relations. Once it has reached the region, a thing or material can be transformed and moved numerous times. When looking at the circular movement of things in Qaanaaq, it becomes clear that people in Qaanaaq, although part of a global market economy through fishing, supply and consumption of oil and other goods, are placed in a particular and marginal position within this economy, a capitalist logic of value, that is, a logic in which commodities define the value system (Tsing 2013: 21) is not dominant among inhabitants in Qaanaaq. Rather, as Nuttall argues, often there is a clash between different logics of exchange and circulation: 'one small-scale, emphasising kinship, reciprocity and cultural identity, the other market-oriented within a context of nationbuilding' (Nuttall 2000: 35). These logics are not always easy to distinguish, and Story 1 shows how they can become entwined. While the Greenland government aspires for its economy to be more integrated into the global economy, and hence welcomes mining, commercial fishing, trophy hunting, among other economic activities, hunters and households have other concerns and priorities that partly overlap and partly diverge from aspirations of economic growth and integration into the global economy. I return to these in Story 2.

As in other places in the Arctic (see Collings 2010; Harder and Wenzel 2012), in Qaanaaq, the sharing and distribution of meat from catches, as well as other resources at hand, form a crucial fabric for non-capitalist relations (Nuttall 2000; Hendriksen and Hoffmann 2014). When the first narwhal of the year is caught, in April or May, the hunter who harpooned it will usually not trade its *mattak* to the factory or down the coast, but keep a good share to distribute among close family and kin, and sell the rest within the community, to some of the many inhabitants who are longing for fresh *mattak* after a long winter without it. A hunter who sells the entire *mattak* of a narwhal caught early in the season to the local factory may not be esteemed or spoken well of in the community. Such events articulate a morality connected to a long tradition of sharing and distributing meat from catches in a network of relations (see Nuttall 2000; Gearheard et al. 2013). The relations and practices that unfold around hunting and other productive activities in Qaanaaq are part of a moral economy of sharing, which at times enters into tension with monetary modes of engaging with things and meat.

One question growing out of Story 1 is how this local economy is (dis)connected to national and global markets, and how this has changed during the last decades, with intensified globalisation and climate change. Another question is how resources get translated in novel ways and converted to cash or new kinds of products because of climate change. The next stories touch upon these questions.

## STORY 2: 'ADAPTABLE' LIVES. SUBTLE DISAGREEMENTS ON THE 'RESOURCES' TO BE TAPPED

In January 2016, Greenland's minister of hunting, fishing and farming visited Qaanaaq, with a delegation from the Greenland government, to conduct public meetings, and to inform the community about a possible zinc- and lead-mining project on the northeastern coast of Greenland, many kilometres away. At a public workshop focusing on business and employment in the Qaanaaq district, the minister gave a speech in which he emphasised the uniqueness of the district due to its breathtaking scenery and the good taste of its products, especially the narwhal *mattak* that is much enjoyed by settlements and towns in all of Western Greenland. He further stressed the challenges of operating and sustaining commercial business in these northern latitudes: the lack of seafood buyers and the logistic conditions causing an increase of the cost of goods.[3] The minister pointed out the 'untapped business opportunities' present in the region, especially the intensification of halibut fishing and trophy hunting, that is, selling of hunting licences to foreign visitors. Lamenting that the few permissions given to operate trophy hunting (on muskox and caribou) in the district are not being used, he emphasised that this particular activity could 'to some extent help improve the economy of the local area ...'. The minister told the hunters that he considered them to be 'adaptable' to fishery, and that developing these untapped opportunities would benefit not only the district, but the entire Greenlandic economy.

With Story 2 we situate Qaanaaq within a national Greenlandic and global economy, by following the minister's lines of the argument, as well as the flow – and restriction on flows – of animals. It is well established that the animals that migrate to and inhabit Avanersuaq have for millennia provided the basis of the subsistence and the livelihoods of *Inughuit*, the people who live in Qaanaaq and the surrounding area (Grønnow and Sørensen 2006; Hastrup 2015). In his speech, the minister mentioned a few of these animals: narwhal, halibut, muskox, caribou and *possibly* polar bear. He made it clear that the value of these animals was to be understood in terms of income-generating activities, as potential resources: how much money, employment and business could be generated around them.

Commercial halibut fishing is new in the region; the warming of the oceans has brought this species of fish to such northern latitudes (Hastrup 2018). The first halibut catch was purchased by a local factory in Qaanaaq in 2010. This factory, the only one in town, was then run by Arctic Green Food A/S; a private Greenlandic company buying and processing fish and other food products in towns and settlements along the Greenlandic west coast. When Arctic Green Food went bankrupt in 2013, people from Qaanaaq organised and founded Inughuit Food. That very same year, almost 110 tonnes of halibut were sold to the factory in Qaanaaq, beating the record of previous years. Halibut was exported to national and international consumers as far away as the Far East, optimism rose and Qaanaaq was celebrated as Greenland's youngest fishing town. In 2014, Inughuit Food entered into partnership with Greenland's biggest company, Royal Greenland A/S,[4] to form Inughuit Seafood A/S, with 50 per cent ownership each. Since then, Inughuit Seafood A/S has been running the factory, steadily increasing its turnover. In 2017, a new record was reached: the factory bought 225 tonnes of halibut from the town's 50 or so fishermen. While many still identify more as hunters than fishermen, halibut fishing *is* an opportunity for monetary income that hunters in the district are tapping into, and an increasing number of households depend on and live well from the income from halibut fishing.

If asking the hunters what untapped opportunities exist in their area, the answer might differ from that of the minister. To hunters and their families, halibut and narwhal are not the only animals of significance when it comes to securing their livelihoods. All animals count, and walrus, seal, narwhal and polar bear are among the most important, for subsistence and also as potential 'cash species'.

With global warming, the season with solid sea ice on which hunters depend for travelling, hunting and halibut fishing has become some four months shorter (Hastrup 2018; Flora et al. 2018). Likewise, the period with open water has become longer, meaning good boats and outboard motors, along with petrol, are increasingly necessary to reach hunting grounds where the animals can be caught. While the traditional dog sledges are crafted locally, the acquisition of boats, motors and fuel depends on a cash economy. When registered as fishermen, people can access significant subsidies from the government of Greenland for buying boats and motors. In this way, the benefits and monetary income from fishing become an infrastructure that supports subsistence hunting (Harder and Wenzel 2012). Still, many hunters lament not being able to hunt more polar bears, walrus and narwhal. Since 2006, hunting of these animals has been regulated by the government of Greenland, restricted by annual quotas, established to secure that hunting is sustainable for the animal population (Wiig et al. 2014; Andersen et al. 2018). Another strong motivation for restricting the hunting of such sea mammals is the sensitivity of global consumers of other species caught by fishermen further down the coast. If news of Greenlandic hunters catching too many polar bears or belugas reached sensitive consumers in other parts of the world, this might trigger an immediate halt to the export of cod, shrimps and Atlantic mackerel; living resources caught far beyond the limits of Avanersuaq hunting grounds. With the recently acquired status of the polar bear as the 'emblem of the new threats of the climate' (Hastrup 2009: 258), the risk of these emotio-environmental chain reactions connected to particular animals becomes higher. Greenlandic decision-makers are thus constantly afraid of the international repercussions a change in quotas on vulnerable and nearly threatened animals might have. These concerns are not groundless: they can be seen as a result of previous bans on Arctic products. Until the 1990s, the international market for sealskin was favourable, and hunters throughout Greenland made a good income from its sale. As an effect of global animal rights groups campaigning heavily against commercial seal hunting in Arctic Canada, and influencing European trade bans on seal products, the trade value of sealskin dropped significantly, and now only represents a minor income to hunting households. In this way, globally connected occurrences have consequences for the environmental management schemes to which hunters in places like Avanersuaq have to adapt.

Due to these market conditions and regulatory regimes, two main 'cash species' remain for hunters to 'tap' and sell in Qaanaaq: halibut that is caught from the sea ice from late December through May, and *mattak* from narwhal that can be caught from May through July or September, and traded to the factory until its storage capacity of 200 tonnes has been reached. It frequently occurs, however, that one of the factory freezers breaks down, and the trade of halibut is put on hold due to diminished freezing capacity. With one weekly plane it can take weeks for the right mechanics to reach Qaanaaq and the installations to be repaired. On such occasions, fishers' and hunters' monetary income is brought to a halt for weeks; the lack of human resources at hand and the limited transportation infrastructure slow down the accumulation and circulation of products and market-driven money flows.

According to the annual quotas and the catch reporting system, approximately 83 walruses, 105 narwhals – some with tusk, some without – and 24 polar bears are annually caught in the Avanersuaq district. Besides the meat they provide for subsistence of humans and dogs, and for sharing and selling, these animals come with added value in their fur, ivory, tusk, skull and claws. Hunters can sell these to tourists, visitors or temporary foreign employees; or the by-products can be transformed into handicrafts or jewellery, which are also sold. However, since the mid-2000s, tight restrictions and bans on the export of these products have been introduced, following international conventions on the protection of endangered species (CITES).[5] As a result, the potential conversion of these by-products from catches into a flow of money has been reduced significantly, making it harder to make a living as a hunter (Nuttall 2016: 362–3).

In sum, the adaptability required of hunters and their families is restricted as much by national and international stakeholders as by changes in the climate and sea ice, and shifting seasonalities of the animals available. Hunters sometimes claim that restrictions are not necessary since climate change and *hila* (weather, the universe) protect the animals, by making them less accessible due to ice or weather conditions (Gearheard et al. 2013). Restrictions also make it harder to adapt. As one hunter told us, ten months before leaving Qaanaaq to get a degree as an Arctic guide: 'We have always been good at bending our lives according to changes in the environment. But now, hunting laws and restrictions prevent us from bending.'

In response to the minister's speech in Story 2, one might ask: adaptable – but to what changes and whose priorities? And urgencies defined by whom? Similar findings and arguments have been presented by Mark Nuttall, who – based on research carried out in communities south of Qaanaaq – argues that climatic changes must be understood in the context of and in relation to other societal changes (Nuttall 2016).

Hunters in Avanersuaq and the Greenlandic government disagree on the framing of problems and urgencies that they need to respond to: the main priority of the Greenland government is to improve the weak national economy, by intensifying fishery and tourism, as well as the extraction of minerals; these are all efforts that support the project of nation-building (Bjørst 2016; Nuttall 2016). The priorities of hunters and their families are different, however, and partly overlap and partly diverge from the government's aspirations for economic growth and integration into the global economy. The priorities of hunters and households are to have a voice in how to lead their lives between homes and hunting grounds, and ultimately, to have their knowledge of the environment and animals count when political decisions are made.

## STORY 3: CATCHING TOURISTS AND HARVESTING SCIENTISTS

'In January I catch fish, in April I catch tourists.' A hunter in his mid-fifties said this as we were crossing the fjord on his dog sledge, returning to Qaanaaq after placing boats and fuel for an upcoming walrus-tagging mission, in which we were participating, along with five hunters and one biologist from the Greenlandic Institute of Natural Resources. The hunter, renowned for his skills, has since the 1980s regularly worked for biologists, serving as collaborator and facilitator for scientific endeavours in the region. Biologists are able to count, calculate and model animal populations in standardised ways, but they depend on the skills of hunters to navigate the sea and icescape, and to reach and manipulate the animals. In return for this work, hunters receive an attractive daily wage (Andersen et al. 2018). From April to August, scientists as well as tourists make up a particular population of living resources to be harvested, or tapped, by hunters.

It is evident in the media and richly documented in the academic literature that 'the melting ice in the Arctic has become an icon of global climate change along with the polar bear clinging to the last tiny ice-floe in the sea' (Hastrup 2009: 245).

With news of climate change and the intensified melting of the Arctic circulating globally, Qaanaaq attracts wealthy tourists, reporters and film crews that arrive to witness and document the unique and melting landscape, and to experience and portray what they see as the last real Arctic hunters. From April to June or early July, when the sun has returned and there is still ice on the fjord to carry dog sledges to and from the spectacular hunting site of the ice edge, and to and from neighbouring settlements, the six small double rooms of the hotel in Qaanaaq, as well as other limited housing options in town, are fully booked. Every week, between five and ten dog sledges depart from Qaanaaq, each with a visitor as passenger, a hunter as guide, and 9 to 13 dogs pulling the party. For such trips, the hunters' association has fixed a price per tourist per day on sledge. While the monetary income a hunter can make in the season of tourism and scientific fieldworks is handsome, this income has to be accompanied by income from hunting. Only if the income from hunting makes up at least 50 per cent of the total annual income of a vocational hunter will his hunting licence be renewed (Piniarneq 2016: 8). On paper, tourists or scientists do not count as game and thus do not count as income from hunting; nor do walrus, polar bear, seal, muskox or caribou, whose meat is not traded through the factory, but enters informal networks of sharing, or goes to feed dogs. It is ironic that while hunters are tapping into the opportunity of selling their skills to a growing influx of visitors, they are deprived of the right to be what tourists from the whole world find attractive about them: hunters of the High Arctic.

The changing sea and land ice regimes have made dog sledge travel routes disappear, and opened up for new seaways that will permit movements that connect places in new ways. These might create new routes for tourists and other visitors to reach Qaanaaq, and generate a circulation of capital, a demand for services in a tourism sector, and potential for further commodification of traditional livelihoods and practices. In turn, hunters and biologists, with differing interests, are concerned about the impact of increasing sea traffic on the animals and ecosystems of the region. Hunters wonder where the animals will move, and how such activities will transform their hunting areas, while biologists are concerned about how narwhals, colonies of breeding birds, walruses and other species will respond to the increased traffic in the area, and the increased potential of oil spills.

Story 3 shows how the circulation of tourists, science and globalising environmental regimes recalibrates aspects of hunting livelihoods in Qaanaaq, and how the logic of wildlife management gets tangled up with that of environmentalism, and those subsisting and surviving in an environment and society that are undergoing rapid change. It shows how livelihoods, skills and materials can be converted into new commodities with economic value, and how social relations are accommodated to these (Bear et al. 2015). With climatic changes bringing new people to Qaanaaq, activities and livelihoods are changed, and so are the relations around them. With the influx of tourists and scientists, hunting livelihoods can be converted into a tradeable product or service. The globalised economy and environmental changes generate new opportunities but also new modes of relating. People in Qaanaaq are active in making the particular conversions and translations, of things, services, tradeable goods and the social relations connected to them.

## ADAPTATION. BENDING ... BUT TO WHAT?

The three stories narrated from Qaanaaq contain points of intersection that can be framed and understood in terms of a changing global economy and climate change. However, the trajectories of the heterogeneous materials narrated in the stories are different in scale. Thule-style mitts and whips, polar bear fur and walrus ivory are not sold via eBay

China but move through networks of non-capitalist social relations, demand and supply. What, then, is to be learned about climate change and the global economy, and how they intersect, when bringing these stories together? What do we learn about changing life conditions in the High Arctic?

The ability to adapt or to bend modes of living to changing circumstances appears in all stories, and is articulated by people in various positions. But what is being adapted, what is adaptable, and who are the subjects of adaptation (individuals, communities, authorities)? The answer differs when asking a hunter or a public employee in Qaanaaq, or a minister from the capital. The versions might be connected or overlapping, coinciding in climatic changes causing loss of sea ice, warming temperatures, and deep concerns regarding future livelihoods. The suggested responses to changes, however, and the urgency behind those responses, vary. People positioned differently, geographically or economically, respond to changes, adapt or bend in different ways. And these different ways may only partially be about, or connected to, globalisation or climate change. We follow Kirsten Hastrup (2013b) in arguing that it matters how we as scholars 'scale our attention' when trying to understand these changes, and how we connect these different ways to globalisation, capitalism and climate change in our analysis and writing. This scaling must be done with attention to how the concerns and urgencies are articulated by the people living in the sites we engage with, and with attention to the fact that none of the changed practices can readily be translated to adaptation to either climate change or the global economy. The three stories show that practices and livelihoods are transformed in and through already existing networks where social relations, income possibilities, education, skills, knowledge, technologies and materials at hand are all constituents.

Kirsten Hastrup has described the ability of people in the High Arctic to adapt as 'flexibility' (Hastrup 2009). Building on Gregory Bateson, she writes that 'flexibility has been the hallmark of local resilience in a world of flux, as has the sense of community within which one knew who one was' (Hastrup 2009: 267). This point is helpful when seeking to situate urgencies, concerns and responses, and understand their directionality and reach. When Jensine moves goods up and down the coast, one purpose is to sell products and generate money. But as she travels and trades along the way, she establishes and maintains relations, and identifies herself, to others and to herself, as well rooted in the place where she was born, and as a knowledgeable person celebrating birthdays, Halloween and up-to-date on how to buy fashionable wallpaper. When a hunter 'catches tourists' and takes them along on his dog sledge through the land that is his, the catch and income is new, but the activity and technology used is recognisable. The network of these relations may extend further than they did three generations ago. They move through Facebook, websites, family ties, eBay, up and down the coast, into people's living rooms and dog sledges, and out towards the world. Through trading practices, fishing and tourism, people in Qaanaaq accommodate and scale concerns and practices in ways in which they can recognise themselves. Some things, however, like national hunting policies and global environmental conservation sensibilities, are more difficult to accommodate or connect to existing livelihoods than others.

## CONCLUSION

Like the Arctic landscape, where all elements are in constant movement, and mobility – or bending – along with these moving elements has been a strategy for human survival for millennia, a strategy for anthropology when seeking to understand globalisation and climate change might be to situate our stories of urgencies and concerns between and beyond settled categories and standardised modes of understanding. One implication of this is that we question how different concerns about climate change and an overheated world, across positions, urgencies and interests, can actually be shared?

This chapter has deliberately broken down what we call 'monster concepts' – global climate change and global capitalism – to puzzling pieces consisting of particular practices, concerns and urgencies, which are linked to livelihoods in Qaanaaq, and situated these in economic practices that are at once local and global. The pieces should make it clear that it is not possible to see or assess climate change or the global economy as a whole from a single perspective. Rather, practices, changes and their description are always situated. If anthropology is to provide missing pieces, we suggest, then, that these pieces should not be seen as parts of a puzzle to be completed, but rather as pieces that do the job of puzzling and unsettling categories and understandings of fixed scale and generalisable scope. Through the stories, we have discussed different versions of adaptation to a changing or 'overheated' world (Eriksen 2016). We argue that in Qaanaaq, climate change concerns partially overlap with economic concerns (related to local livelihoods), which are, in turn, connected to the globalisation of markets and economy, as well as national and international politics. These partially connected forces, and *the way* they connect, sometimes exacerbate each other. How they do so is specific to the way capital moves to and from Qaanaaq, and to the specific ways in which resources are valued, used, commercialised – by people positioned differently economically and geographically, and to how the environment is changing. In this nexus, adaptation means to bend livelihoods, practices and ways of knowing to economic as well as climatic changes, in forms very specific to the High Arctic. Adaptation can thus never be understood as happening in a linear, straightforward and singular modality. Rather, it depends upon who is adapting to what, with what means, and on what scale. It follows that anthropology's scholarly urgency lies in studying and reporting on such complex processes in all their specificity and complexity.

## ACKNOWLEDGEMENTS

This chapter draws from research carried out in Northwest Greenland between 2014 and 2017 under the interdisciplinary 'North Water' (NOW) project, which was generously funded by the Carlsberg Foundation and the Velux Foundations.

Thanks to all colleagues and participants in the NOW project, especially to Kirsten Hastrup, with whom fieldwork experiences and thinking were shared. Thanks also to Astrid B. Stensrud and Thomas Hylland Eriksen for organising the 'Climate change and capitalism' workshop and for engaged and thorough editorial work, and also thanks to colleagues at Aalborg University for productive feedback on an earlier draft of this chapter.

## NOTES

1. All names have been changed.
2. Fieldwork was carried out in April–July and October–November 2015 and May–June 2016, as part of the interdisciplinary NOW Project that studied living resources and human societies in a long-term perspective.
3. http://naalakkersuisut.gl/da/Naalakkersuisut/Nyheder/2016/01/210116-Qaanaaq-havomraader (accessed 1 February 2019).
4. Royal Greenland A/S is the largest company in Greenland; a fishing, production and export company fully owned by the Greenland government.
5. CITES (the Convention on International Trade in Endangered Species of Wild Fauna and Flora) is an international agreement between governments aiming to ensure that international trade in specimens of wild animals and plants does not threaten their survival.

## REFERENCES

Adger, N., N.W. Arnell and E.L. Tompkins. 2005. Successful Adaptation to Climate Change Across Scales. *Global Environmental Change* 15: 77–85.

Andersen, A.O., M.P. Heide-Jørgensen and J. Flora. 2018. Is Sustainable Resource Utilisation a Relevant Concept in Avanersuaq? The Walrus Case. *Ambio* 47(S2): 265–80.

Appelt, M. 2004. De Sidste Dorsetfolk. In H.C.Gulløv (ed.), *Grønlands Forhistorie*. Copenhagen: Gyldendal, pp. 170–200.

Baer, H.A. 2012. *Global Capitalism and Climate Change: The Need for an Alternative World System*. Walnut Creek, LA: AltaMira Press.

Bear, L., K. Ho, A. Tsing and S. Yanagisako. 2015. Gens: A Feminist Manifesto for the Study of Capitalism. Theorizing the Contemporary, 30 March. Available at: https://culanth.org/fieldsights/652-gens-a-feminist-manifesto-for-the-study-of-capitalism (accessed 1 February 2019).

Bjørst, L.R. 2016. Uranium – The Road to 'Economic Self-sustainability for Greenland'?: Changing Uranium-positions in Greenlandic Politics. In G. Fondahl and G.N. Wilson (eds), *Northern Sustainabilities: Understanding and Addressing Change in the Circumpolar World*. Cham, Switzerland: Springer International Publishing, pp. 25–34.

Bohannan, P. 1959. The Impact of Money on an African Subsistence Economy. *Journal of Economic History* 19(4): 491–503.

Collings, P. 2010. Economic Strategies, Community, and Food Networks in Ulukhaktok, NT, Canada. *Arctic* 64(2): 207–19.

Eriksen, T.H. 2016. Overheating: the World since 1991. *History and Anthropology* 27(5): 469–87.

Flora, J. 2019. *Wandering Spirits. Loneliness and Longing in Greenland*. Chicago; London: University of Chicago Press.

Flora, J., K.L. Johansen, B. Grønnow, A.O. Andersen and A. Mosbech. 2018. Present and Past Dynamics of Inughuit Resource Spaces. *Ambio* 47(S2): 244–64.

Gearheard, S.F., L.K. Holm, H. Huntington, J.M. Leavitt, A.R. Mahoney, M. Opie, T. Oshima and J. Sanguya. 2013. *The Meaning of Ice. People and Sea Ice in Three Arctic Communities*. New England: International Polar Institute Press.

Grønnow, B. 2016. Living at a High Arctic Polynya: Inughuit Settlement and Subsistence Around the North Water During the Thule Station Period, 1910–53. *Arctic* 69(5): 1–15. doi: http://dx.doi.org/10.14430/arctic4573.

Grønnow, B. and M. Sørensen. 2006. Palaeo-Eskimo Migrations into Greenland: The Canadian Connection. In J. Arneborg and B. Grønnow (eds), Dynamics of Northern Societies: Proceedings of the SILA/NABO Conference on Arctic and North Atlantic Archaeology, May 2004, Copenhagen. *National Museum Studies in Archaeology & History*, Vol. 10. Copenhagen, pp. 59–74.

Gudeman, S. 2001. *The Anthropology of Economy*. Oxford: Blackwell.

Harder, M.T. and G. Wenzel. 2012. Inuit Subsistence, Social Economy and Food Security in Clyde River, Nunavut. *Arctic* 65(3): 305–18.

Hastrup, K. 2009. Arctic Hunters: Climate Variability and Social Flexibility. In K. Hastrup (ed.), *The Question of Resilience. Social Responses to Climate Change*. Copenhagen: The Royal Danish Academy of Sciences and Letters, pp. 245–70.

—— 2013a. Anticipation on Thin Ice: Diagrammatic Reasoning in the High Arctic. In K. Hastrup and M. Skrydstrup (eds), *The Social Life of Climate Change Models: Anticipating Nature*. London and New York: Routledge, pp. 77–99.

—— 2013b. Scales of Attention in Fieldwork: Global Connections and Local Concerns in the Arctic. *Ethnography* 14(2): 145–64.

—— 2015. The North Water: Life on the Ice Edge in the High Arctic. In K. Hastrup and F. Hastrup (eds), *Waterworlds: Anthropology in Fluid Environments*. London and New York: Berghahn Books, pp. 279–99.

—— 2018. A History of Climate Change: Inughuit Responses to Changing Ice Conditions in North-West Greenland. *Climatic Change* 151(1): 67–78. doi: 10.1007/s10584-016-1628-y.

Hastrup, K., A.O. Andersen, B. Grønnow and M.P. Heide-Jørgensen. 2018. Life Around the North Water Ecosystem: Natural and Social Drivers of Change Over a Millennium. *Ambio* 47(S2): 213–25.

Hendriksen, K. and B. Hoffmann. 2014. Qaanaaq Distrikt – Infrastruktur og Erhvervsgrundlag: Sammenfatning af Empiriindsamling i Qaanaaq. Center for Arktisk Teknologi, Sisimiut.

Kopytoff, I. 1986. The Cultural Biography of Things. In A. Appadurai (ed.), *The Social Life of Things: Commodities in Cultural Perspective*. Cambridge: Cambridge University Press, pp. 64–93.

Nissen, M, L.T. Pedersen, S.M. Olsen, M.B. Kreiner, G. Dybkjær, R.T. Tonboe, J.L. Høyer, S. Ribeiro and N. Mikkelsen. 2017. *Report on the Temporal and Spatial Evolution of Sea Ice in Inglefield Bredning, and the Relationship between Ice Properties and Hunting Trips*. ICE-ARC, Deliverable Report D3.21.

Nuttall, M. 2000. Choosing Kin: Sharing and Subsistence in a Greenlandic Hunting Community. In P.P. Schweitzer (ed.), *Dividends of Kinship: Meanings and Uses of Social Relatedness*. London: Routledge, pp. 33–60.

—— 2016. Narwhal Hunters, Seismic Surveys, and the Middle Ice: Monitoring Environmental Change in Greenland's Melville Bay. In S.A. Crate and M. Nuttall (eds), *Anthropology and Climate Change, From Actions to Transformations*. New York and London: Routledge, pp. 354–72.

Parry, J.P. and M. Bloch (eds). 1989. *Money and the Morality of Exchange*. Cambridge: Cambridge University Press.Piniarneq. 2016. Jagtinformation og Fangstregistering. Available at: www.businessingreenland.gl/~/media/Fiskeri%200g%20fangst/Fangst%200g%20jagt/PINIARNEQ_2015/Piniarneq%202016%20DA_ENG.pdf (accessed 1 February 2019).

Polanyi, K. 1957. The Economy as Instituted Process. In K. Polanyi, C.M. Arensberg and H.W. Pearson (eds), *Trade and Market in the Early Empires: Economies in History and Theory*. Glencoe, IL: Free Press, pp. 243–70.

Strathern, M. 2004. *Partial Connections*. Walnut Creek, LA: AltaMira Press. Originally published 1991.

Taussig, M. 1977. The Genesis of Capitalism Amongst a South American Peasantry: Devil's Labor and the Baptism of Money. *Comparative Studies in Society and History* 19(2): 130–55.

Tsing, A. 2013. Sorting Out Commodities. How Capitalist Value is Made Through Gifts. *HAU: Journal of Ethnographic Theory* 3(1): 21–43.

Wenzel, G. 2000. Sharing, Money, and Modern Inuit Subsistence: Obligation and Reciprocity at Clyde River, Nunavut. In G.W. Wenzel, G. Hovelsrud-Broda and N. Kishigami (eds), *The Social Economy of Sharing. Resource Allocation and Modern Hunter-Gatherers*. Osaka: National Museum of Ethnology, Senri Ethnological Studies No. 53, pp. 61–85.

Wiig, Ø., E.W. Born and R.E.A. Steward. 2014. Management of Atlantic Walrus (*Odobenus rosmarus rosmarus*) in the Arctic Atlantic. *NAMMCO Scientific Publications* [S.l.], 9: 315–41. doi: http://dx.doi.org/10.7557/3.2855.

# 8. Counting
## Health Emergencies and the Constitution of Extractive Natures in Northern Loreto, Peru

*María A. Guzmán-Gallegos*

INTRODUCTION[1]

In 2015, I interviewed José Alvarez, the Director of Biodiversity of Peru's Ministry of Environment under the Humala government (2011–16) about oil extraction in Northern Loreto in the Peruvian Amazonia. He recalled a visit he made to Vista Alegre, a village located in the Upper Tigre River, while he was working at the Augustinian Catholic Mission 20 years earlier:

> It was 1994. I arrived at the village and its leader took me to its cemetery. I will never forget the sight of the two rows of tiny, white-painted crosses made of rustic wood. In less than two years, 21 children had died with their stomachs swollen up and spewing out pieces of their liver. A few years earlier, in 1986, nine children had died with similar symptoms. People did not know why this was happening. Some believed that so much death was due to the evil acts of shamans while others related it to the continuous oil and chemical leakages coming from an oil production site, Bartra Producción, operated by the American oil company Occidental Petroleum [OXY], near Vista Alegre. Many decided to move out believing that too many were dying, attempting in this way to save their children's and their own lives. Even so, some of them died years after moving out. For a small community like Vista Alegre, this high death toll meant the loss of almost an entire generation.

Vista Alegre, like many other native communities within the areas of oil extraction in Northern Loreto, was greatly affected by oil spills in the 1980s and 1990s.[2] These were the years of civil war in Peru, social mobilisation was not allowed and those who engaged in protests could be accused of being terrorists. Neither local leaders nor even missionaries or teachers, I was told, dared to complain. People's concerns about recurrent ailments and the ongoing environmental changes were dismissed and often actively silenced. When Vista Alegre's local authorities and the village's teacher went to Iquitos to inform the authorities about the continuous deaths of children, and to demand the visit of a health brigade, they got no response from the regional health authorities or from OXY.

This chapter deals with contamination produced by oil extraction in Block 1AB/192 in Northern Loreto, Peru.[3] Focusing on the Upper Tigre River, one of the main rivers crossing this oil field, it explores how people's recurrent ailments, which were previously ignored, have become matters of public concern and state action during the last ten years. This region has historically been a frontier tied to cyclical booms in different commodities; a frontier characterised by racialised and ethnic structures of exclusion and by the violent depletion of forests and the biophysical degradation of ecosystems (Hennessy 1978: 12; Little 2001: 1). In Northern Loreto, oil has been one of the main commodities for the past 45 years. Lack of regulation and control institutions characterised its extraction in the first 20 years; a new boom in the extraction of oil and deregulation has marked the last two decades. By 2015, oil concessions covered 46 per cent of Loreto and 75 per cent of the Peruvian Amazonian region. While environmental legislation was passed and health and environmental control institutions were established in the first decade of the twenty-first century, this boom has been followed by new legal reforms that narrowed the scope of former legislation and weakened the control institutions' functions. The last wave of such reforms occurred in 2014 and 2015, when several laws and decrees were passed. These laws (Law 30320 and Law 30327) weakened environmental requirements, and limited the control capacities of environmental state institutions. According to the Humala government, the explicit goal of these legal reforms was to secure foreign investment in mining and oil extraction in Peru, which could be affected by falling prices in global markets.

The issues dealt with in this chapter resonate with those discussed in the vast literature on environment and health and on environmental justice. Broadly speaking, this literature has discussed the disputes over knowledge regarding the health effects of exposure to chemically defined pollution and industrial waste and the issue of uncertainty often asserted by states and corporations (Mitchell and Cambriosio 1997; Murphy 2000, 2004; Fortun 2004; Kirsch 2004). These studies analyse the endurance of toxic substances, often not accounted for, as well as the distribution of environmental harm, which disproportionally affects indigenous and dispossessed communities. The longevity of toxic materials and the disproportionate burden for some populations has been discussed in terms of 'slow violence', a violence which is dispersed across time and space, and which contributes in effect to constitute some populations and places as suitable to be sacrificed (Kuletz 1998; Nixon 2011; Bohme 2014; Voyles 2015).

In recent works on mining and oil extraction, contested knowledge continues to be the subject of analysis. Seeking to obtain licences to operate or to manage conflict, corporations employ strategically scientific knowledge to disengage from the environmental and health effects extraction may create (Kirsch 2014; Li 2015). Scientific knowledge is deployed, it is suggested, by the corporations and the state to create a discourse on scientifically based management of risk and accountability. The latter presupposes establishing commensurabilities and equivalences (Li 2016), which are central to compensation politics.

In this chapter, I build upon the insights of these works, and explore how establishing different types of commensurabilities and equivalences reinforces the constitution of what I call 'extractive natures', that is 'natures' that are embedded in and the product of extraction practices. The latter include the particular ways of extracting oil, that is, the social and material conditions under which extraction occurs. The constitution of 'extractive natures' happens, I argue, in spite of ongoing social mobilisation that aims to question oil extraction in Northern Loreto. An important premise for the concept 'extractive natures' and for my analysis is the understanding and analytical stance that nature is not an entity, a physical reality existing 'out there', independently of particular socio-material and political conditions and practices. Nature is, in other words also the product of particular conditions and actions. This is also why I use 'natures'.

In my discussion of the ways in which commensurabilities and equivalences enter into the constitution of 'extractive natures' I draw on the works of Nelson (2010, 2013, 2015) and Verran (2010), in particular, their analysis on the double condition of counting and of numbers. Counting, Nelson (2015) suggests, entails both quantification (enumeration) *and* recognition (what matters), that is, what is counted and counts, who is counted and counts and, also, what is considered to be worth repairing. Quantification and recognition are central to the politics of compensation and to the governance of extraction. Both build upon a logic of balancing out, that is, of quantifying losses and benefits, of making equivalences, in ways that they are meant to balance out, that is, to reach a zero. I discuss how counting, through the quantification of oil-related substances has contributed to making pollution coming from oil extraction real, and to establish a possible relation between the existence of pollutants in soils and water bodies and human populations' wellbeing and health. The possibility of such a relation has made people's *current* experiences of oil extraction, and their lives, visible and relevant.

Numbers, however, do more. Depending on the relational forms numbers take in enumerating, they may contribute to constitute natural entities, such as rivers or river waters. According to Verran, numbers materialise the relation unity/plurality in different ways by taking either the one/many form or the whole/parts form. She characterises the one/many form as 'potentially containing unity within the plurality of many, and whole/parts as having plurality contained within unity' (2010: 173). While numbers in the one/many form serve to establish value, numbers in the whole/parts form establish order. Numbers can serve to audit the quality and the state of river water or soils, and they can serve, simultaneously, to constitute 'natural comprising wholes'. An example of a 'natural comprising whole' which I shall discuss later in this chapter, is 'Norwegian Nature', as Asdal explains in her work (2008, 2011). Such a 'natural whole', she argues, was brought into being in Norway in the 1980s as part of political negotiations about acid rains in Europe, and through heterogeneous material and knowledge practices. The object of political concern and analysis changed from particular rivers to Norwegian Nature, an entity that came to comprise all rivers in Norway. Another example of a natural whole, constituted through quantification, is the Australian water market, which Verran describes and I shall discuss later.

By using the concept 'extractive natures', I want to shift the focus in the analysis from natural resources that are extracted to the socio-natural entities which are created. In the constitution of extractive natures, numbers serve, on the one hand, to establish the environmental condition of soils and water bodies, and to determine the value of what has been lost. On the other hand, I suggest that numbers serve to reinforce a constant focus on parts (parts of a river, of forests, of a population) and on immediate harm. This focus on parts and immediate harm negates the long-term effects that extraction and contamination have on people's bodies and health, and reinforces the extractive practices that continuously sacrifice excluded populations.

## THE UPPER TIGRE RIVER:
## MOBILE POPULATIONS AND OIL INFRASTRUCTURES

In the early 1970s, the Peruvian government granted a concession to OXY for oil exploration in Block 1AB/192, located close to the borderline between Ecuador and Peru. These borderlands had been disputed areas since the first decades of the nineteenth century, and until the 1940s people travelled back and forth over the borders between Ecuador and Peru. Most of the Kichwa inhabitants of the Upper Tigre River trace their origins to Kichwa villages located along the Bobonaza River, the Upper Curaray, the Conambo and the Pastaza, in what is now Ecuadorian territory. In these borderlands, they moved in search of new places, founded villages and were part of extensive intermarriage and trade networks with neighbouring Achuar groups. Villages were also founded close to rubber patrons' farms. Movement and migration patterns changed radically when the war between Ecuador and Peru broke out in 1941 and the borderline was redefined 300 km to the north. During the 1940s and early 1950s, the Peruvian military, often combining physical violence and patronage relations, relocated dispersed Kichwa families into villages as a strategy to gain control over these areas and their inhabitants.

Displacement, relocation and migration waves re-emerged in the early 1970s when oil extraction started. The first encounters between the inhabitants of the Upper Tigre River and the exploratory companies subcontracted by OXY led to a series of involuntary and voluntary displacements. Since the Peruvian state considered these lands and forests as uninhabited, families had to leave their homes. Hoping to get access to electricity and health care, new villages were founded near oil wells and processing plants. The functioning of oil installations led moreover to the temporary migration of oil workers. Since OXY did not employ local people, oil workers came from Iquitos or from other Peruvian cities. The presence of oil workers and the construction of an oil refinery around 1975-77 also prompted an increasing migration of Spanish-speaking traders who started small businesses in these villages.

While oil extraction has lasted for more than four decades, oil concessions, installations and corporations have changed throughout the years, producing in turn new movements and transformations. Oil production that started in 1972 was intensified by the 1980s. OXY extended its installations on the Corrientes River, and increased refining and producing activities in the Tigre River, turning Block 1AB into Peru's largest oil field in terms of production and infrastructure. Wells, production facilities, refineries, minor oil pipelines, air bases, ports and an extensive road network connecting these facilities were built. Years later, a large part of this infrastructure was closed down; of almost 250 drilled oil wells, 76 have been temporarily abandoned while 40 are permanently closed.[4] In the mid-1980s, OXY closed down a refinery located near Marsella, including the harbour and all adjacent roads, and moved its installations upstream; many families left Marsella and founded four new villages. Local mobility increased as a result of changes in corporate practices. Pluspetrol Norte, which in 2000 acquired further rights to Block 1AB, started employing the villagers. From that year to date, men from all the different villages located within Block 192 move temporarily to work on its different installations. Relocation and displacement had also been common in times of illness and death, as happened in Vista Alegre. People from the affected areas moved south to the lower part of the Tigre River, east to the Napo River or to other provinces and regions.

Oil infrastructure has also changed. As Murphy (2013: 3) suggests, in her work on chemicals' toxicity, oil infrastructure comprises not just spatially fixed installations such as wells, pipelines and refineries, but also chemicals, metals and hydrocarbons that 'become mobile, travel in water ways, settle in landscapes, are absorbed by bodies and bio-accumulate in food chains permeating in many and different ways the lives of humans and non-humans'. Oil leakages occurred regularly, turning the waters of the Tigre River greasy and black for several days. Leakages came from oil wells and production facilities on the upper part of the river, and from the cargo ships that for several years transported oil from Marsella to Saramurillo on the Marañon River and to the Loreto region's capital, Iquitos. Many of these ships were in poor condition and frequently collided with each other in the many bends of the river. Regardless of where these leakages occurred, oil spills travelled downstream, entering the small lakes connected to the Tigre River.

In summer time, when the river flow dropped significantly, the Tigre River's water used to become greenish and extremely salty. Men were not able to obtain clean water to drink while hunting, since streams that flow into the Tigre River could also be salty. A common corporate practice, which lasted until 2009, was indeed to dispose of formation water[5] directly into the Tigre River, its tributaries and adjacent lakes and streams (Yusta-García et al. 2017). Although its chemical composition varies, formation water is toxic and at least four times saltier than ocean water (E&P Forum/UNEP 1997). Thinners used to make oil flow through the pipelines were also disposed of in water bodies close to the production facilities.

People also experienced the changed conditions of water bodies when they went fishing. Antonia,[6] a woman living in Vista Alegre, told me that her husband and sons used to find dead fish in the Montano Lake, which is close to the Bartra production facility and to Vista Alegre. It was not just the dead fish, she said, they noticed changes in the bodies of the fish, for instance, the scales of the doncella fish (*Pseudoplatystoma tigrinum*) became thicker and when grilled the whole fish became hard; its meat was not tasty and its liver had scars as if somebody had burned it with a cigarette. As some fish went from the lake to the river, it was not easy to avoid catching fish that they knew might have a bad taste.

Oil exploration and production activities also changed the routes of animals and hunters. While the noise produced by helicopters, boats and the various machines found in the production facilities led many wild animals to change their routes away from the forests near the oil installations and the villages, salty waters attracted others. Salty streams and defectively closed wells, the latter found scattered within the Block 1AB/192 area, attracted tapirs, which can still be found licking the salty waters that leak from closed wells. As tapirs changed their routes so did hunters, who include wells in their hunting tracks (see also Orta-Martínez et al. 2018).

Oil compounds also travelled through the air, affecting insects and plant life. People from Marsella still remember when clouds of black smoke covered the sky for several days. They knew then that OXY's workers were burning what people called in Spanish the *afrecho del petróleo*, 'the oil bran'. The 'oil bran' referred to oil that had leaked from the oil installations in the Marsella refinery and was collected in a small stream, named *Gringo yaku* (literally Gringo water) that runs on its borders. When it was filled up with thick layers of oil, usually every third month, they set 'the oil bran' on fire; oil residues could then burn for several days. Delicate plants such as papaya trees or cocona bushes dried up as a result. Bothersome insects and mosquitoes disappeared for several years.

In the first three decades of oil production, these landscape transformations and environmental degradations were often ignored, as were the ways these changes permeated the lives of human and non-human inhabitants of the Upper River Tigre. Even in cases where human death tolls were inexplicably high, such as in Vista Alegre, neither health authorities nor corporations took action. Local realities came to count only when contamination emerged as an environmental condition.

## MAKING CONTAMINATION REAL

Norberto, a man in his sixties, and I were enjoying the sight of a clear stream which crosses the village of Paiche Playa where his grandchildren were bathing. Suddenly he asked me if I had noticed that the Paiche Playa people were heavy. Laughing at my surprise, he told me jokingly that it is because their bodies are filled up with lead. Adopting a more serious tone, he said that people until recently did not know that oil leakages could have dangerous health effects. Whenever there were oil spills close to Paiche Playa, people used to collect oil residues and dry them, to make candles. Children and adults got sick, but it is always difficult to know what causes illness, he concluded. As for Norberto, oil contamination, the multiple changes in water bodies and soils, and the effects these may have on the animals people eat and on the people themselves have become an issue of increasing concern for the inhabitants of the Upper Tigre River. This concern is related to the emergence of contamination as an environmental condition, a condition that may have health effects and that requires action.

Such emergence depends on the production of particular types of knowledge and on specific legal, political, institutional and material configurations and practices (Asdal 2008, 2011; Sawyer 2015). This section focuses on some of these configurations, exploring in particular the work of numbers. I will discuss how the act of counting helps to establish the presence of oil-related components as contamination, and how it makes people's experiences and ailments relevant.

As mentioned above, Vista Alegre's high death toll in the 1990s did not lead to state or corporate action. What caused (and still causes) people to become sick and eventually die was, as Norberto also points out, uncertain and unknown. Several reports produced by the Regulatory Body for Energy Investment and the National Environmental Health Agency stated, however, that surface water samples showed high concentrations of hydrocarbons (oil and fat), barium, lead and chlorides in the Tigre, Corrientes and Pastaza Rivers and in several of their tributary streams. Studies conducted by the Research Institute of the Peruvian Amazon (IIAP) showed that water bodies had concentrations of lead and copper that were above the permissible limits defined in Peruvian Water Law. Concentration of these metals in the tissues of fish from rivers and lakes in the area indicated bioaccumulation of these compounds, making fish meat unfit for human consumption (Orta-Martínez et al. 2007: 5–7). Although the linkages between health and chemical exposures are characterised by uncertainty (Murphy 2004), neither the health state authorities nor OXY made any effort to connect available information with the deaths in Vista Alegre. According to Tomas Villalobos, at that time in charge of the Augustinian Mission, OXY's physicians claimed that it was a yellow fever epidemic, and vaccinated its workers. Worried about the fact that some villages were significantly more affected than others, Villalobos together with José Alvarez, whose narration opens this chapter, contacted a physician in Lima, who was in charge of surveying the proliferation of tropical diseases in Peru (see also La Torre 1998). He suggested that such deaths could be caused by the combination of acute poisoning and latent hepatitis B or latent liver problems. The deaths coincided indeed with a period of leakages of toxic thinners into the Montano Lake where the people of Vista Alegre used to fish.

It was, however, not until 2005 that linkages between pollution and health were made explicit. A study of water quality and the human health situation in the upper part of the Corriente River was carried out in 2005, as a response to increasing social mobilisation. The study showed that 66.21 per cent of blood samples from children under 18 years of age exceeded the maximum allowable limits of lead according to the US Centers for Disease Control and Prevention level of 10llg/dl, and that 98 per cent of them had cadmium blood levels above the allowable limit of 0.1llg/dl; 99.2 per cent of blood samples from adults tested exceeded the limit for lead, while 64.8 per cent of blood samples from adults exceeded the biological tolerance level for cadmium (Lu 2009: 31). Eventually, after several uprisings, the indigenous organisation of the Upper Rio Corrientes reached an agreement on behalf of the 30 communities it represented with state representatives and Pluspetrol,[7] according to which Pluspetrol had to stop disposing of production water in water bodies, clean open pits, and initiate a health programme for the Corrientes River (Bebbington and Scurrah 2013). Oil-related contamination in other basins, such as the Tigre and Pastaza Rivers, and their inhabitants' ailments, were not taken into account.

Since oil spills and leakages continued in the Upper Tigre, the Upper Corrientes and the Upper Pastaza Rivers, a new process of social mobilisation started in 2011. The indigenous organisations leading it demanded that the National Environmental Health Agency (DIGESA), the Water National Authority (ANA), the Agency for Environmental Assessment and Control (OEFA) and the Regulatory Body for Energy Investment and Mining (OSINERGMIN) conduct environmental monitoring in the Tigre, Corrientes and Pastaza Rivers. The reports that were produced between 2012 and 2014 revealed – once again – the presence of heavy metals such as lead, barium and mercury, of chemicals such as total nitrogen and ammonia nitrogen, and of total petrol hydrocarbons (TPH) and polycyclic aromatic hydrocarbons (PAH) in water bodies and soils, especially near oil installations. The amount of pollutants exceeded Peruvian regulations and other standards such as the Dutch Pollutant Standards and those of the World Health Organization (WHO). As a result, in 2014 the Upper Tigre River, the Upper Pastaza River and the Upper Corrientes River were declared to be in a state of health and environmental emergency.

Although the quantification of oil pollutants and how it relates to thresholds on permissible levels is difficult to grasp for most people, numbers are important in the villages, as the following illustrates. After having visited an abandoned and leaking well, Miguel Larco and Santiago Rosero, from Remanente and Vista Alegre, and I were reviewing the reports that DIGESA, OSINERGMIN and OEFA had published in 2013. Miguel and Santiago worked at the communities' environmental monitoring system; they had participated in taking samples of soils and water bodies, as did several villagers, but they had not seen the final reports. We identified in the diagrams some water bodies located nearby and found the indicated quantities of different pollutants. Neither they nor I understood if some of the amounts were above or below the standards and thresholds used in the report. When we found that they were above the thresholds, Miguel and Santiago showed great relief, which first seemed paradoxical to me. However, they explained to me that people from Vista Alegre and other villages nearby, such San Juan de Baltra, located close to the Baltra Production site, had seen their loved ones die. They all used to go fishing in the Montano Lake where they on several occasions found dead fish. It would be unfair if the numbers in the reports did not show this, since people knew that the quality of the lake's water was not good; it would mean that their suffering was not taken into account. Hence, the numbers acknowledged their suffering.

Contrary to what had happened in Vista Alegre in the 1990s, when no action was taken in spite of high death tolls, the counting and measuring of pollutants served to assess the quality of water. Comparing these amounts with standards (national or international), the numbers defined whether water at particular places was toxic and inappropriate for human consumption. In this way, the reports suggested possible connections between polluted water and health, which hopefully could be confirmed by blood samples. For the villagers, this meant that their experience of contamination and their suffering were made visible and relevant.

As Nelson (2013: 19) explains, permissible limits are, however, not a representation of an actual observation but rather the product of calculation in highly controlled conditions. Indeed, limits and standards measure levels of chemicals (usually) in water and in human bodies and replicable responses, the latter based on common physiological reactions to each chemical. Such limits and standards, for example, WHO's guidelines, assume particular bodies and conditions of exposure: an adult male worker exposed eight hours a day, five days a week. The guidelines do not reveal the possible effects a compound of several chemicals may have on malnourished bodies that have been exposed to chemicals in utero. In spite of this, the counting of pollutants, standards and thresholds serves to make people in the communities feel that their lives and their dead counted. To paraphrase Nelson (2015: 43), their lives were made ontologically equivalent to every other human life, and in principle subject to equal protection as humans.

The ways of establishing equivalences changed radically, however, when responsibilities and compensation measures were to be decided. What was recognised as effects of contamination, which places and whose lives were deemed as valuable became contentious issues, as demonstrated in the next section.

## EXTRACTIVE NATURES AND PRESENT HARM

The declaration of a health and environmental emergency led to different and contested agreements with the state and corporations. In spite of the differences between them, the villagers, the media and state representatives perceived most of the agreements mainly as dealing with a debt the corporations and the state owed the villagers for all the harm done. In this sense, these agreements were understood and framed as compensation. The focus on compensation is common in the negotiation of conflicts regarding extraction in Peru; compensation paves the way for mining and oil corporations to obtain social legitimacy for their operations (Arellano Yanguas 2011; Bebbington 2011; De Echave 2018). As Nelson (2010) suggests, compensation presupposes establishing equivalences together with the logic of balancing out, which is common to double-entry bookkeeping practices. Double-entry bookkeeping operates with two columns known as either cost and benefit, debt and credit or profit and loss. The amounts indicated in these two columns ideally have to be balanced out, that is, reach zero. These double-entry bookkeeping practices entail, in the context of extraction in Peru, agreeing amounts of money or securing state investments in health, sanitation or education to compensate for the harm done. Further, the logic of balancing out is constitutive for what I call 'extractive natures'. 'Extractive natures' refers first to the 'natural' entities constituted through the measurement of environmental qualities. Second, it alludes to the entities created through a constant focus on delimited parts (of a river, of forests, of a population) and on immediate harm. As parts of these 'extractive natures', villages, villagers, animals and aquatic life, even water, and oil infrastructure are framed as immobile and 'fixed', as we shall see in the following discussion.

Pluspetrol's community advisers reached different compensation agreements with different villages. They negotiated the amount of money according to the village's bargaining capacity, which greatly depended on its location in relation to current production facilities. In the case of the Upper Rio Tigre, the villages located closer to the facilities received more money than those located further away. The community of 12 de Octubre, close to the Shiwiyacu producing wells received US$750,000 while Marsella, where the leakages from a refinery that closed in 1985 still affect a stream flowing into the Tigre River, received US$90,000. The people of Vista Alegre were offered US$1,800, which they rejected. The villagers explained to me that Pluspetrol offered this money as compensation for the destroyed *aguajal*, a field of *aguaje* palms affected by a defectively sealed oil well. An engineer from the evaluation company, employed by Pluspetrol, had simply counted the palms and decided an amount, but, as the villagers pointed out, she did not take into account all the monkeys and birds that no longer came to the *aguajal* because the palms no longer bear fruit. The engineer did not count the deformed and tasteless fish that the villagers caught in the Montano Lake and ate. Even worse, in the accounting, the engineer and Pluspetrol ignored the deaths: all the people who still die with swollen stomachs, vomiting blood.

Since the villagers of Vista Alegre, like most of the villagers in the Upper Tigre River, considered that the oil corporations and the state were indebted to them for the harm they had done, they challenged not just the amount of compensation but also what was included and what had been valued. The palms could not be separated from the monkeys that depend on the palms' fruits, just as humans could not be considered apart from the fish they eat. They challenged, moreover, a particular 'oil temporality' that denies possible connections with past and future times.

The limits of this temporality and of what or whose lives were valued came to the fore and were questioned when documentation on the effects of heavy metals and hydrocarbons on human health was produced. The biologists and physicians, who were supporting the indigenous leadership, were afraid of not being able to establish the certainty of these linkages. Wanting to avoid a situation in which the state and the corporations denied responsibility, as they have done before, the scientists considered it necessary to take blood samples solely from villagers currently living near oil-producing installations. The need for certainty resulted, however, in ignoring people's continuous mobility. Those inhabitants of the Upper Tigre River who had been working throughout the years in different oil installations, in remediation of abandoned sites and in cleaning oil spills without proper protection were excluded. By focusing on the present time, and on current harm, it was not considered that oil infrastructures are, as chemical infrastructures, temporally extensive and uneven. Some chemicals break down quickly while others endure, injuring organisms slowly (Murphy 2013: 3). Measuring the presence of heavy metals in blood samples gathered mainly in villages affected by current oil production ran the risk of not accounting for the slowness, persistence and latency of certain chemicals and hydrocarbons.

The focus on limited losses and on current damage is closely related to the constitution of particular 'extractive natures'. Such a constitution presupposes a particular framing and ordering of wholes and parts. Asdal's (2008, 2011) and Verran's (2010) discussion on how numbers and heterogeneous quantification practices contribute to the emergence of 'nature-wholes' is illustrative in this regard. In 'Enacting things through numbers: Taking nature into account/ing', and in *Politikkens Natur – Naturens Politikk,* Asdal discusses how Norwegian Nature as a natural entity, as a 'nature-whole', emerged in Norwegian politics. She argues that Norwegian Nature as an entity existing out there, as an object to be protected and governed, did not exist a priori, but was brought about. Norwegian Nature as a nature-whole emerged as part of the politics and negotiations about acid rain in Europe, and was 'the result of nature-objects collected, made present, visible and imaginable through a set of heterogeneous material practices' (Asdal 2008: 126). Numbers, counting and quantification were crucial in this respect. Indeed, during the 1970s, the focus was on how acid rain affected particular rivers, and on fish stocks in these rivers. In other words, particular rivers were the object of intervention. In the 1980s, this focus changed gradually when the notion of critical load was established – that is, the quantity of pollution 'nature' can tolerate without damage – and when state budgets allowed water researchers to compile samples from over 1,000 rivers which could have been affected by sulphur and nitrogen compounds. *This* or *that* river were transformed into Norwegian water surfaces and thus into Norwegian Nature. The notion of critical load, the collection of analysed water samples, the reports written and the maps which showed the affected rivers allowed for the constitution of 'a nature-whole' and for making 'Norwegian Nature' real and visible: an entity that should be taken into account. Rivers became parts of this whole (Asdal 2008: 126–7).

Numbers and quantification practices may bring about various kinds of 'nature-wholes'. In her work, Verran explores how the monitoring of water quality of Australian rivers, creeks and dams serves, on the one hand, to represent aquatic life and assess the failing health of Australian rivers. This monitoring occurs through the extensive counting of tiny creatures living in water bodies. Through the aggregation of these numbers, a general picture of the quality of the Australian waters emerges. On the other hand, these numbers also serve to constitute quite another whole – the Australian water market. The water market is constituted as 'articulable water resources' which can be separated into many parts such as surface water, ground water, return water or reuse water.

Each part is made separable as a product that can be sold, a commodity (Verran 2010: 176–7).

In Peru's neoliberal political and economic configurations, the ways numbers are used, and the work numbers do, reinforce current oil extraction practices and the socio-natural configurations they create. I argue that the quantification of pollutants in several locations in Northern Loreto, the search for certainty in the relation between these pollutants and the presence of metals in human blood samples *together* with establishing compensation sums lead to the constitution of 'extractive natures'. These are not nature-wholes to be protected, but rather constituted in ways that make them 'extractable'. The constitution of extractive natures, I suggest, is first of all characterised by the continuous focus on the parts that make up the whole. Second, the whole to which parts belong is obscured by defining it as irrelevant.

As mentioned, numbers make manifest the relation between unity/plurality either by taking the one/many form or the whole/parts form. In the work done to quantify the existence of pollutants, which again reveals the quality of water bodies and soils within Block 1AB/192, numbers take the one/many form. When taking many water and soil samples, counting and measuring are crucial in establishing the existence of many milligrams of heavy metals, or many heavy metals in many water bodies, in many soils. Adding up these numbers and comparing them with different standards of permissible levels make visible the environmental quality of water and soils, a quality that the authorities can no longer ignore. At the same time, as described above, water and soil samples are taken in particular places along a river, especially in those places where oil installations are or have been located. The whole which is made temporarily relevant is the area – the forests, rivers, villages and all oil infrastructure – of what *currently* is Block 1AB/192. However, constituting contaminated parts of a river does not entail that an entire river, even less a river basin, is constituted as a whole. In other words, not even a river is constituted as a visible and relevant nature-whole. Further, the attempts to establish a particular certainty between the existence of pollutants and health, and the compensation measures that favour some villages, reinforce this continuous constitution of parts. Thus, even Block 1AB/192 as a whole vanishes. The block's parts which are seen as relevant are those where oil-producing installations are located (Figure 8.1).

Indeed, when the counting and enumeration of pollutants took place between 2005 and 2010, the focus was not on the whole Corrientes River, but only a part consisting of five communities on the Upper Corrientes River. In 2014, increasing social mobilisation led to the inclusion of the Upper Pastaza and the Upper Tigre Rivers. The focus was and still is

*Figure 8.1* The continuous constitution of (river) parts. 'You are beyond [the company's] area of influence. If the river dragged the oil it is not our fault'.

Source: http://archivo.larepublica.pe/carlincaturas/carlincatura-2015-02-05 (accessed 30 April 2017.

mainly on some villages in the upper part of the rivers and on the people who happened to be living there at a particular time. Moreover, only the immediate present is relevant; past grievances or future health risks are not taken into consideration.

## CONCLUSION

Through the counting and quantification of pollutants in rivers and soils in Northern Loreto, oil contamination has emerged as an existing environmental condition, and people's experiences and ailments have been acknowledged by state institutions. Counting and quantification have made it possible to question the conditions of oil production in Block 1AB/192. At the same time, numbers serve to reinforce the constitution of 'extractive natures' due to the continuous focus on certain localities, together with the need to assert the certainty of the linkages between contamination and health effects. Separated parts of water bodies, of habitats, of people are framed, while wholes – for instance, forest habitats, rivers, mobile populations or even the oil concessions – are not. This spatial focus on parts is reinforced by a temporal focus on current damages, on the basis of which compensation is agreed. Oil contamination is dealt with as an emergency event and not as an intrinsic part of oil production in Northern Loreto. This also means that neither the longevity of oil pollutants nor the long-term health effects this might have on human and non-human bodies are taken into account.

## NOTES

1. This chapter is based on six months of fieldwork carried out in 2015 and 2016 in four communities of the Upper Tigre River. In addition to participant observation in daily activities, in community meetings and visits to contaminated sites, I conducted informal conversations and 150 semi-structured interviews with leaders of indigenous organisations and communities, and with representatives of non-governmental organisations (NGOs) and state institutions. In addition, I draw on my participation in activities and meetings related to local environmental monitoring while I was working in the area from 2007 to 2013.
2. 'Native communities' refers to the communities that the Peruvian state acknowledges as indigenous. I use native communities as it is also the term used by the communities themselves where I worked.
3. An oil field is the area given in concession to an oil company field. Oil fields in Peru are called oil blocks. In this chapter I use this designation. 1AB was renamed Block 192 in 2014. For the sake of simplicity, I use Block 1AB/192.
4. Numbers vary. According to Martínez Orta et al. (2007), 400 producing oil wells have been drilled and of these over 200 have been abandoned.
5. Formation water is a layer of water found in reservoirs of oil; formation water comes up when oil or gas is extracted.
6. All personal names are fictitious.
7. As mentioned, Pluspetrol bought OXY's rights to oil block 1AB in 2000.

## REFERENCES

Arellano Yanguas. 2011. *Minería sin Fronteras? Conflicto y Desarrollo en Regiones Mineras del Peru*. Lima: Pontifica Universedad Católica del Peru.
Asdal, K. 2008. Enacting Things Through Numbers: Taking Nature into Account/ing. *Geoforum* 39: 123–32.
Asdal K. 2011, *Politikkens Natur – Naturens Politikk*. Oslo: Universitetsforlaget.
Bebbington, A. 2011. *Minería, Movimientos Sociales y Respuestas Campesinas: Una Ecología Política de Transformaciones Territoriales*. Lima: IEP-CEPES.
Bebbington, A. and M. Scurrah. 2013. Hydrocarbon Conflicts and Indigenous Peoples in the Peruvian Amazon: Mobilization and Negotiation Along the Rio Corrientes. In A. Bebbington and J. Bury (eds), *Subterranean Struggles, New Dynamics of Mining, Oil and Gas in Latin America*. Austin, TX: University of Texas Press, pp. 173–96.
Bohme, S.R. 2014. *Toxic Injustice*. Oakland, CA: University of California Press.
De Echave, J. 2018. *Diez Años de Minería en el Perú*. Lima: Cooper Acción.

E&P Forum/UNEP. 1997. Environmental Management in Oil and Gas Exploration and Production. An Overview of Issues and Management Approaches, *UNEP IE/PAC Technical Report 37 or E&P Forum Report 2.72/254.*

Fortun, K. 2004. From Bhopal to the Information of Environmentalism: Risk Communication in Historical Perspective. In G. Mitman, M. Murhpy and Ch. Sellers (eds), *Landscapes of Exposure, Knowledge and Illness in Modern Environments*. Chicago, IL: University of Chicago Press, Osiris, Second Series, Vol. 19, pp. 283–97.

Hennessy, A. 1978. *The Frontier in Latin American History*. Bristol: Edward Arnold.

Kirsch, S. 2004. Harold Knapp and the Geography of Normal Controversy: Radioiodine in the Historical Environment. In G. Mitman, M. Murphy and Ch. Sellers (eds), *Landscapes of Exposure, Knowledge and Illness in Modern Environments*. Chicago, IL: University of Chicago Press, Osiris, Second Series, Vol. 19, pp. 167–82.

—— 2014. *Mining Capitalism: The Relationship Between Corporations and Their Critics*. Oakland, CA: University of California Press.

Kuletz, V. 1998. *Tainted Desert: Environmental and Social Ruin in the American West*. New York: Routledge.

La Torre, L. 1998. Sólo Queremos Vivir en Paz. *IWGIA* 25, Copenhague.

Li, F. 2015. *Unearthing Conflict, Corporate Mining, Activism and Expertise in Peru*. Durham, NC: Duke University Press.

—— 2016. In Defense of Water: Modern Mining, Grassroots Movements, and Corporate Strategies in Peru. *The Journal of Latin American and Caribbean Anthropology* 21(1): 109–29.

Little, P. 2001. *Amazonia, Territorial Struggles on Perennial Frontiers*. Baltimore, MD and London: Johns Hopkins University Press.

Lu, M. 2009. *The Rio Corrientes Case: Indigenous People's Mobilization in Response to Oil Development in the Peruvian Amazon*. Master's thesis, International Studies Department, University of Oregon.

Mitchell, L. and A. Cambriosio. 1997. The Invisible Topography of Power: Electromagnetic Fields, Bodies and the Environment. *Social Studies of Science* 27: 221–71.

Murphy, M. 2000. The Elsewhere Within Here: Or How to Build Yourself a Body in Safe Space. *Configuration* 8: 87–120.

—— 2004. Uncertain Exposures and the Privilege of Imperception: Activist Scientists and Race at the U.S Environmental Protection Agency. In G. Mitman, M. Murhpy and Ch. Sellers (eds), *Landscapes of Exposure, Knowledge and Illness in Modern Environments*. Chicago, IL: University of Chicago Press, Osiris, Second Series, Vol. 19, pp. 266–82.

—— 2013. Distributed Reproduction, Chemical Violence, and Latency. *Scholar Feminist Online*. Available at: http://sfonline.barnard.edu/life-un-ltd-feminism-bioscience-race/distributed-reproduction-chemical-violence-and-latency/ (accessed 23 March 2016).

Nelson, D. 2010. Reckoning the Aftermath of War in Guatemala. *Anthropological Theory* 10(1–2): 87–95.

—— 2013. 'Yes to Life = No to Mining' Counting as Biotechnology in Life (Ltd) Guatemala. *Scholar Feminist Online*. Available at: http://sfonline.barnard.edu/life-un-ltd-feminism-bioscience-race/yes-to-life-no-to-mining-counting-as-biotechnology-in-life-ltd-guatemala/0/?print=true (accessed 19 February 2016).

—— 2015. *Who Counts? The Mathematics of Death and Life After Genocide*. Durham, NC: Duke University Press.

Nixon, R. 2011. *Slow Violence and the Environmentalism of the Poor*. Cambridge, MA: Harvard University Press.

Orta-Martínez, M., D.A. Napolitano, G.J. Maclennan, C. Ocallaghan, S. Ciborowski and X. Fabregas. 2007. Impacts of Petroleum Activities for the Achuar People of the Peruvian Amazon: Summary of Existing Evidence and Research Gaps. *Environmental Research Letters* 2(4): 045006. doi:10.1088/1748-9326/2/4/045006.

Orta-Martínez, M., A. Rosell-Melé, M. Cartró-Sabaté, C. O'Callaghan-Gordo, N. Moraleda-Cibrián and P. Mayor. 2018. First Evidences of Amazonian Wildlife Feeding on Petroleum-contaminated Soils: A New Exposure Route to Petrogenic Compounds? *Environmental Research* 160: 514–17.

Sawyer, S. 2015. Crude Contamination: Law, Science and Indeterminacy in Ecuador and Beyond. In H. Appel, A. Mason and M. Watts (eds), *Subterranean Estates, Life Worlds of Oil and Gas*. Ithaca, NY and London: Cornell University Press, pp. 126–46.

Verran, H. 2010. Numbers as an Inventive Frontier in Knowing and Working Australia's Water Resources. *Anthropological Theory* 10(12): 171–8.

Voyles, T. 2015. *Wastelanding, Legacies of Uranium Mining in Navajo Country*. Minneapolis, MN and London: University of Minnesota Press.

Yusta-García, R., M. Orta-Martínez, P. Mayor, C. González-Crespo and A. Rosell-Melé. 2017. Water Contamination from Oil Extraction Activities in Northern Peruvian Amazonian Rivers. *Environmental Pollution* 225: 370–80.

# 9. Expansive Capitalism, Climate Change and Global Climate Mitigation Regimes
A Triple Burden on Forest Peoples in the Global South

*Harold Wilhite and Cecilia G. Salinas*

### INTRODUCTION

In this chapter we will argue that people who live in forests in the 'peripheries' – as defined from a Western-centrist perspective – are suffering a triple burden brought on by (1) extractive capitalism; (2) climate change, resulting from the high-energy consumption practices of societies at the 'centre'; (3) and more recently from market-based global environmental regimes established in the name of climate mitigation. Beginning in the colonial period, foreign-based commercial enterprises have intruded on forest ecosystems everywhere around the globe. In the populated forests of the global South, this has disturbed the lives of millions of people. Pressures on forests and the people living in them increased in the early stages of capitalist expansion. Over the past decade, both transnational corporations and local industries have accelerated the extraction of forest products for national and global markets. Parallel to these intrusions with their sources in the various manifestations of the globalising economy are policies made in the name of environmental protection, such as the establishment of national parks and science-based forest management. These environmental regimes have added to pressures on forest people's livelihoods and in some cases resulted in their eviction from their home territories. More recently, environmental programmes created to combat climate change, such the United Nations programme, *Reducing Emissions from Deforestation and Forest Degradation in Developing Countries* (UN REDD), and the Clean Development Mechanism (CDM) treat forests as carbon sinks, reducing the rich complexity of human-nature interactions to market exchanges. In this chapter we will draw attention to UN REDD, a system of payments to forest communities in exchange for preserving biomass. The evidence thus far shows that the commodification and marketisation of forests in the name of climate mitigation perturbs the lives of forest peoples rather than bringing improvements. Based on experiences thus far, we claim that UN REDD's neglect of the fullness of human-forest interaction has negative impacts on forest livelihoods and fails to cope with the main source of deforestation, the practices of extractive industries and agribusiness.

At the heart of the problem in both globalising capitalism and global environmental regimes is a fallacious representation of human-nature relationships. Humans are viewed as apart from, or exogenous to, the ecosystems in which they live. From a capitalist perspective, nature is conceived as a repository of resources to be converted to commodities and the people living in nature are posited as either obstructions to resource extraction or potential labourers for extractive industries. Research on nature conservation reveals that there are cases in which programmes ostensibly attempt to respond to the critique of a narrow understanding of the human-nature relationship by discursively aiming to integrate people with the environment. However, these efforts are inadequate, partly because the policies and the perceptions that inform them are based on a strong dose of scientific arrogance, dismissing the importance of local subsistence livelihoods and local knowledge on how to live in balance with ecosystems. Unfortunately, this remains at the heart of global climate mitigation regimes, making people and their forest-based activities the problem, diverting attention from the main source of carbon emissions, namely, high material and energy consumption in the rich countries. As Cabello and Gilbertson write, in the thinking behind global climate policy, there is 'little space for discussing how to start transforming the very structures (market thinking) that created this problem (climate change) in the first place' (2012: 164).

It is important to stress that these globalising forces (capitalism and environmental regimes) affect millions of people living in tropical forests who depend on them for food, animal fodder, fuel and building materials. According to the World Commission on Forests and Sustainable Development, 350 million of the world's poorest people, among them 60 million indigenous people, depend almost entirely for their subsistence and survival on forests. In the analysis of the triple burden on forest-dependent peoples, we include communities inside forests, whose basic life sustenance is dependent on shifting cultivation and/or hunting and gathering. In many parts of Asia, Africa and Latin America, forest people are indigenous people or minority ethnic groups. Also included in this broad category are those belonging to small rural communities with a mixed ethnic background and a mixed micro-economy based on subsistence agriculture and a cash economy.

In this chapter, we briefly discuss the effects of the consolidation and expansion of global capitalism on forest-dependent peoples, beginning with its origins in colonialism. We follow this with a discussion of how the global climate mitigation regime UN REDD extends and reinforces many of the same principles and pressures of economic and environmental intrusion. We will highlight the paradoxes involved in making Southern forests the locus of climate mitigation projects, showing how mitigation regimes are grounded in the same capitalist logic that justified defining forests as natural capital and disturbed the livelihoods of forest-dependent peoples.

## FORESTS AS RESOURCE-REPOSITORIES FOR ECONOMIC DEVELOPMENT

Colonialism in the Americas, Africa and Asia fostered global trade and a growing class of merchants who accumulated massive amounts of wealth. This was accompanied by a growth in the influence of intellectuals who exalted reason and argued that societal progress was dependent on scientific knowledge. Not only were local subsistence practices in the colonies discounted as being non-modern and inefficient, but also forest extractivism was established as a 'natural' form of human-nature interaction. These notions were imported into early capitalism and facilitated the extension of globalising markets into the global South in the nineteenth century. The accompanying market thinking stripped nature of its intrinsic value, ignoring fragile interrelationships within nature and reducing it to a repository of resources to be harvested and exchanged in global markets. First colonial and then capitalist industries established themselves in the forests of the South, where raw material such as minerals and timber fed into the production of manufactured goods for global markets. According to Acosta (2011: 66), 'In practice, extractivism has been a mechanism of colonial and neo-colonial plunder and appropriation ... of the raw materials essential for the industrial development and prosperity of the global North.' Logging is one of the oldest and most intrusive of the extractive industries. In recent times the growing global pulp industry has been responsible for the deforestation of large tracts of rainforests in the global South (Carrere and Lohmann 1996) as well as the replacement of natural forests with forest monocultures in the form of tree plantations (see Tsing 2005; Salinas 2011; Schiavoni and Alberti 2014). Other industries with extensive forest-based operations include agribusiness, livestock production and mining.

An important target for mining extraction is coal, the most plentiful deposits of which lie buried under old-growth forests. In the words of Naomi Klein, 'coal was the black ink in which the story of modern capitalism was written' (2014: 175). Forests were not only disturbed due to mining extraction but, as Acosta (2011) points out, in order to be productive, mining operations must use large quantities of toxic chemicals and vast amounts of water. The mining of coal produces enormous quantities of non-recyclable waste. The coal-rich forests of India, China, Indonesia, Bolivia and Tanzania are examples of Southern forests that have been and continue to be targeted by extractive industries, many of them with ties to national and transnational corporations. The extracted coal is burned to produce the energy that fuels capitalism, but it also produces carbon dioxide – the main contributor to climate change. The bulk of carbon emissions emanate from the production of electricity that is used in industrial production, and increasingly over the past century, to fuel energy-intensive lifestyles in the wealthy countries of the world. Carbon dioxide is absorbed by biomass and the oceans, but the absorptive capacity of the oceans is reaching its limit. Excess $CO_2$ migrates into the troposphere, where it impedes the dissipation of heat into and through the atmosphere, resulting in increases in global temperature as well as climate perturbations.

The rapid acceleration of global economic growth in the wealthy parts of the world over the past half century has increased the tempo of resource extraction in the forests of the South. The accumulated assault on forests has exceeded nature's tempo of regeneration in many forest ecosystems, leading to a reduced capacity to absorb $CO_2$ as well as to widespread species extinction and disturbances of the subsistence economies of local communities whose livelihoods depend on forest biomass. According to Kjosavik and Vedeld (2011: 21), capitalist intrusion in local forest economies in the developing world creates 'the contradiction between the needs of expanded reproduction of capital and the needs of daily and intergenerational reproduction of labour power, [disturbing] the biophysical environment through resource demands, waste disposal and pollution, all of which impact on the essential ecological processes and life support systems'. We will return to the discussion of capitalism and forests in the last section of the chapter. First, we will focus on the political and material transformations of forests in India from the colonial period until today. Thereafter, we will take up cases that show how these forest policies, based on Western scientific afforestation principles, have negatively affected the livelihoods of forest dwellers in many other parts of the world.

## FORESTS AND THEIR INHABITANTS AS SUBJECTS OF GLOBAL ECONOMIC AND ENVIRONMENTAL REGIMES

The imposition of laws and regulations on Indian forests and the people living in them in the name of national economic development and environmental protection provide early historical examples of how the livelihoods and interests of forest peoples have been disregarded and disturbed (Wilhite 2014). In the colonial period in India, which stretched from the late 1700s until independence in 1947, the British passed a succession of laws intended to facilitate more efficient extraction of natural resources and forest products in their Indian colony. They established a Forestry Department and enacted forest protection laws in which forest peoples were defined as encroachers. A series of forest acts in the nineteenth century not only denied the rights of forest-dwellers to live in the forests, but also denied them the use of any of its products. The rights, traditional practices and economic base of millions of forest-dwellers were affected, many of whom were tribal peoples (adavasis) who had been living in Indian forests for centuries. Since these people's ideas and practices regarding health, spirituality, kinship, economy and material culture are all tied to the forest ecosystems in which they live, legislating away their access to nature was tantamount to legislating away the very principles on which their communities and livelihoods were based.

These Indian colonial policies were tied to nineteenth-century environmental discourses which conceptualised nature as a repository of resources, to be harnessed, controlled and managed using scientific principles ('scientific forestry'). According to Oosthoek (2007), the aim of scientific forestry was to pursue maximum sustainable yields of timber, and profit, based on quantitative forestry management involving extensive surveys of tree types and timber stocks. Growth rates of tree species were calculated and the fastest-growing species, producing high timber yields, were planted. Wide swathes of forest were reduced to timber plantations. Scientific forestry ignored or rejected the importance of ecosystem balance. There is ample evidence that the implementation of 'scientific' principles in India neither adequately protected forest ecosystems nor contributed anything positive to the forest-dwellers who depended on non-timber biomass (Guha 2006).

In the early twentieth century, large tracts of forest were defined as national parks or wildlife sanctuaries. In the Indian Forest Acts from 1927 until independence, the people living in these forests were defined as illegal encroachers and poachers. A programme of resettlement to more marginal forests was set in motion. Some forest communities refused and engaged in various forms of resistance. In the forests of

Eastern India, many tribal peoples joined armed resistance movements. There is evidence that many of those who were compliant with orders to move never received promised compensation in the form of either cash or land. In many cases, entire villages were moved several times as the passing of more stringent environmental laws designated their new forest as a protected area (Fernandes and Paranjpye 1997).

After independence in 1947, the Indian government continued the scientific and 'sanctuary' forest policies. Through the enactment of the *Forest Policy Resolution* of 1952, the government assumed legal ownership of all Indian forests. Large numbers of people living in forests were forcibly displaced and those who refused to move were denied social services and other citizenship rights. Between 1951 and 1990, the Indian Planning Commission estimates that 20 million people were displaced from their traditional homelands. Of these, 40 per cent, or 8.5 million, were tribal peoples. Only 2 million of these were given adequate compensation in the form of land and/or cash (Kalpavriksh 2002). Over the same period, deforestation driven by mining, logging and tourism increased at a rapid rate.

The landmark Chipco movement of 1973 marked a turning point in how forest peoples were perceived in public opinion and government policy, not only in India but in other parts of the world, due to extensive international media coverage. Faced with the encroachment of logging companies into their traditional forestlands, women from the local community literally hugged trees in order to impede the progress of logging machines. This image of women hugging trees to protect the environment became an icon for environmental protection movements around the world (Guha 2006). In his analysis of the Chipco resistance, Anil Argawal, the journalist who first brought the Chipco story to the Indian national media, pointed out that there was a widespread misunderstanding of the motives of the Chipco protesters: they did not see themselves as indigenous environmentalists who were saving pristine nature, but rather were fiercely protecting their fragile resource base (Argawal and Narain 1985). They understood that expanded logging and other incursions into their forest ecosystem would have catastrophic consequences for their ability to sustain their subsistence-oriented biomass harvesting and agricultural practices. The Chipco movement and the publicity surrounding it led to a reassessment by the Indian government of the myth that forests were better off without people. It fostered new legislation that encouraged cooperation between members of the Indian Forest Service and the people living in the forests. The Joint Forest Management (JFM) Act of 1988 recognised the rights of forest peoples to remain in certain designated forests, where they could harvest limited amounts of non-timber forest products.

In 1990 the Ministry of Environment and Forests took a step further in acknowledging the rights of forest peoples, encouraging Indian states to give a quasi-legal status to 'forest villages'. This was short-lived, however, and was revoked in May of 2002 in a new federal directive demanding the eviction of 'all illegal encroachment of forestlands in various States/Union Territories'. In the 14 months between May 2002 and August 2004, 150,000 families were evicted from forests. Those who refused to leave were denied public services such as schools, health care and electricity. At the same time, the issuing of government permits for commercial logging, for the extraction of plants for research and medicinal purposes, for the building of hydroelectric plants and for the licensing of mining activities increased dramatically in these same forest areas (Wilhite 2014).

Due mainly to pressures generated by media attention on the forced displacement of thousands of forest peoples, in 2006 the Indian government declared that the evictions constituted an 'historical injustice' and Parliament passed the landmark Forest Rights Act (originally entitled the Tribal Rights Act) which 'recognise[d] and vest[ed] the forest rights and occupation of forest land in forest-dwelling scheduled tribes who have been residing in such forests for generations but whose rights could not be recorded' (Indian Government 2006). The Act granted forest-dwellers 'responsibilities and authority for sustainable use, conservation of biodiversity and maintenance of ecological balance'. However, there were contentious parts of the Act, for example, it only applied to families who had lived in their current forest community since 1980, thus denying rights to the millions of people who occupied their current forest land after having been evicted from their former forest homes. Also, in forest areas that have been designated as national parks or wildlife sanctuaries, tribal rights secured under the Forest Rights Act can be 'modified' by the government. In practice, this means that people can remain in the forest, but are moved to the peripheries and not allowed to enter or use large tracts of the forest that are reserved for wild animals. At the same time, a burgeoning safari tourism industry allows vehicles and people to regularly criss-cross these same protected zones.

Studies of nature management in Africa show similar treatment of forest peoples under colonialism and into the independence era. Colonial ideas of nature based in the European Enlightenment's dualism between humans and nature led to local communities' exclusion, to the denial of their land rights and to a view of nature as a resource or as pure and untouched. African nature was to be preserved as pristine, wild and dangerous and the only human presence that was allowed was safari tourism (Newmann 1998; Murombedzi 2003; Nustad 2011). The concept of national parks, invoked under the latter stages of colonial rule, hardened the policies against people living and using the areas incorporated into parks. Access was denied to indigenous communities and nomadic pastoralists, but was open for hunters and nature tourists. As Newmann writes, nature was 'preserved from the effects of human agency by legislatively creating a bounded space for nature controlled by a centralized bureaucratic authority' (1988: 9). The massive Kruger Park in South Africa was one of the first parks designated as a space where tourists could view wild animals in their natural environment. The people who had lived off hunting or pastoral practices in the territory incorporated into the park did not have a place in this vision. Hunting for food or seeking security was redefined as 'poaching'. The herding of domesticated animals through traditional grazing areas was restricted or halted altogether.

The independence of African countries provided new governments with the opportunity to reverse or improve the ways that indigenous peoples were treated in development and environment policies, but as was the case in India, policies hostile to indigenous peoples continued in many African countries after independence (Ellis and Swift 1988). According to Newmann, the creation of a vast game reserve in Tanzania in 1988, the Umba-Mkomazi Reserve, was accompanied by an eviction order affecting 5,000 pastoralists and many other small agricultural communities. This limited their access to ancestral lands and restricted their customary resource use. Around the same time in Mozambique, the grazing lands of many local communities were incorporated into the Mnananara Biosphere Reserve, blocking their capacity to continue their pastoral practices. In Uganda, thousands of people in local communities were evicted from areas that would become the Kibale Forest Reserve and the Queen Elizabeth National Park. The people who had been living in these areas for generations were defined as encroachers. As Newmann writes, traditional forest interactions were made illegal and forest-dependent peoples using biomass or hunting animals for subsistence were defined as criminals guilty of crimes such as 'livestock trespass, illegal hunting, wood theft, and ... species extirpation' (1998: 2).

The history of exploitation and nature management in the Americas from colonial times to the present does not differ much from the cases presented here from India and Africa. Not even the introduction of the category UNESCO Biosphere Reserve as an innovative form of conservation has managed to integrate people with the environment in an organic way. Specifically, the UNESCO Biosphere Reserve Category was launched during the Earth Summit in Rio de Janeiro in 1992, along with the international treaty, the United Nations Framework Convention of Climate Change (UNFCCC), which became the basis of the Kyoto Protocol. This category emerged as a response to the critique directed against protected areas, which, as we have pointed out, separated people from nature, by demonstrating that nature conservation, the sustenance of cultural diversity and development were compatible. Salinas' (2016) study of a Biosphere Reserve in Northern Argentina shows how the conservation practices within the reserve create new separations between people and nature. Indigenous groups within the reserve are not evicted, but the economic interests of large landowners restrict their livelihood practices. Small agricultural communities living in the borders of the reserve are categorised as nature predators who are to be blamed for the destruction of the forest (for similar examples, see Kaus 1993; Martinez-Reyes 2004).

These forest histories from the global South demonstrate that forests have been appropriated and regarded as a 'resource' to be plundered and invaded in the name of economic development and later regulated in the name of environmental protection. The lives of forest-dependent peoples have been invisible in a lens that has regarded forests as bastions of raw materials, biodiversity, wild animals and carbon sinks. In the cases in which they have been acknowledged, forest-dependent peoples have been perceived as impediments to the management of natural resources rather than as sources of knowledge on ecosystem management.

## GLOBAL CLIMATE MITIGATION IN SOUTHERN FORESTS

The report of the World Commission on the Environment (WCED 1987) put climate change on the international political agenda four decades ago. The report indicated that in order to prevent severe climate change, global carbon emissions would need to be reduced by 80 per cent within a half century (the report assigned 60 per cent reductions to the rich countries of the world and 20 per cent to developing countries). It is noteworthy that in the period from 1990 to 2010, global carbon emissions grew by 60 per cent. While the rich countries of the world have barely been able to stabilise $CO_2$ emissions, the emissions from the 'emerging' countries, including China, India and Brazil have grown rapidly. The ability of global society to generate a robust accord limiting carbon emissions has been inhibited by the fact that the sources of climate emissions (coal, oil, natural gas, forest products) are also sources of economic growth, creating disincentives at all levels from global to local for doing something to rein in their use (Wilhite 2016).

The most recent projections from a number of different sources are converging on a likely global temperature rise of in excess of 2 degrees Celsius or more within the next century, which will have major impacts on micro-climates, including severe disturbance of forest ecosystems and potentially devastating effects on food production and food harvesting worldwide (Lynas 2007). The conclusions of the most recent Intergovernmental Panel on Climate Change report (IPCC 2014) include the point that the 'failure to rapidly reduce climate-gas emissions will result in severe, pervasive and irreversible impacts for people and ecosystems'. The extraction of carbon and the consequences of carbon emissions for micro-climates create a double burden for people living in Southern forests. In this section we will argue that the burden is being further increased by the inclusion of Southern forests in global carbon sequestration regimes.

The globally negotiated solutions intended to reduce climate emissions apply the same market logic to emission reduction that drives neoliberal economic theory and practices. It creates a carbon market in which high-emitting countries are awarded credits towards their commitments to reduce carbon emissions in return for investments in renewable energy and reforestation projects in forests of the global South. The two principle global carbon mitigation markets are the Clean Development Mechanism (CDM) and UN REDD. We will focus on UN REDD, which has the most comprehensive interface with forest-dependent peoples. UN REDD is aimed at the sequestration of carbon in forests. The ideas behind REDD were first laid out at the UN climate change negotiations at Bali in 2007. A year later, the UN, with Norway as one of the major backers,[1] jointly launched UN REDD. UN REDD awards carbon reduction credits to high $CO_2$-emitting countries in return for payment to forest communities for increasing forest biomass.

Although the global interest in REDD is growing – as reflected by the number of UN REDD partner countries, which had increased from nine to 64[2] by December 2017 – 'the lack of clarity on the future financing mechanism of REDD+[3] remains a major sticking point for participating countries integrating REDD+ into their national development policies, strategies and plans' (Olding 2017). By 2014 only four of the 340 initiatives have sold carbon credits, and ten received payments conditional on the success of deforestation or degradation measures (Sunderlin et al. 2014). Many of the REDD+ initiatives are supported by bilateral, public or philanthropic funding. The lack of global carbon funding by donors or carbon buyers threatens the feasibility of a stable long-term prospect for the initiative. Due to the uncertain future political and financial landscapes, previous conservation and development strategies are reframed in terms of REDD+. As a result, protected areas that were already protected, or other existing incentives for sustainable management, conservation or sales of certified timber, can apply to receive REDD+ funding, if REDD+ should ever get off the ground (Mathews 2014; Salinas 2016).

There are many difficulties and conflicts in implementing UN REDD, particularly the problems in either ignoring or supporting commercial forest interests. Deforestation in tropical forests can be attributed to infrastructural projects and industrial activities such as mining, hydropower installations, logging and pulp industry, prospecting for medicinal plants, industrial agriculture and livestock grazing. A genuine effort to encourage reforestation would aim at reducing commercial impacts, but instead, because UN REDD acknowledges and rewards all forms of increases in reforestation, including monoculture reforestation, such as coffee, sugar cane and timber plantations, commercial activities are increasing the geographical pressure on the forest territories of indigenous peoples as monoculture plantations extend their reach (Borner et al. 2010).

The evidence from Mexico shows that UN REDD is subsidising commercial activity that is detrimental to forest-dwellers in Mexican forests. In Chiapas, Mexico, payments from UN REDD have been used to support the production of African oil palm and jatropha for biofuels. This not only contributes very little to carbon sequestration, it is adversely affecting the shifting agricultural practices of the indigenous Maya community. According to Osborne (2011), a UN REDD payment of 2,000 pesos was contingent on a community agreement not to use fire to prepare the milpa or to cultivate fallow plots beyond five years of regrowth, both of which are critical to viable milpa-based production. Another study in Chiapas pointed out that UN REDD supports the growth of timber for market exchange but not the growth of fruit trees, traditionally used by the Maya communities to supplement their diets (Diemont et al. 2006). These and other challenges to the livelihoods of marginal forest-dependent peoples are exacerbated by lack of recognition of their land rights. Since historical rights to land are not always clearly designated, UN REDD provides opportunities for private actors who have the economic power to claim and trade land rights (Hecht and Cockburn 2010; Hirsch et al. 2012; Okamoto et al. 2012). 'Carbon cowboys' (Aguilar-Støen 2017) are buying forestland and gaining control over forest areas that later can be offered as part of REDD projects. In New Guinea, where the livelihoods of 87 per cent of the population depend on natural resources for their subsistence needs, Shearman et al.

(2010) found examples in which carbon traders have coerced indigenous peoples into signing their lands over to UN REDD.

UN REDD payments to participating communities are intended to be funnelled through national and local governments in accordance with bilateral agreements. There are many examples of corruption in the handling of UN REDD and other forestation programme payments. For example, in Peru, the Interethnic Association for the Development of the Peruvian Rainforest (AIDESEP) found that the promised payback to local communities in the form of cash for carbon savings has not always reached the communities and individuals affected, due to bureaucracy and corruption (AIDESEP 2010). AIDESEP has also made the point that while the Peruvian state is putting pressure on indigenous peoples to embrace UN REDD, they allow multinational companies to extract timber, and to pollute the Amazon, while exempting them from clean-up. In Tanzania, an evaluation of the early stages of UN REDD revealed significant corruption by the government agency for administering UN REDD, the Ministry of Natural Resources and Tourism (Lang 2013). Carus (2010) reports that in 2010, Liberia demanded the extradition of a British carbon trader on charges of bribery in connection with a deal to lease one-fifth of the country as carbon offsets worth up to US$2.2 billion.

There are no systems in place for monitoring compliance with the UN REDD's system of compensation to forest peoples or the conduct of local authorities with regard to safeguarding international conventions on indigenous rights (Lang 2013). The research of Salinas (2016) in Northern Argentina shows that superficial efforts at monitoring would not suffice, given the lack of transparency in the accounting systems used by local agencies and their tendency to manipulate data in favour of international institutions. Salinas found that facts and information reported to UN REDD are manipulated to show accordance with the requirements of international institutions at the expense of the forest-dependent peoples, both indigenous people and small-scale agricultural communities. Inviting local communities to so-called 'consultation workshops', the promotion of free prior informed consent (FPIC) and the existence of innovative laws and public policies to protect forests and their inhabitants are underscored as symbols of commitment and accountability in reports to international institutions. Salinas found that in practice the implementation is consistently incomplete, unclear and confusing. In reality, the laws are not always enforced, participation can be equated with attendance, and individuals acting on behalf of state institutions and agencies often pursue private interests rather than strictly applying public laws and regulations. In this manner, measuring and monitoring technologies offer no more than an illusion of transparency, which Salinas argues must be best understood as translucency rather than transparency (Salinas 2016). Translucency refers here to a deceptive transparency. The observer perceives things, but is unable to identify with certainty what she is seeing or is deceived into believing that what she sees is representative of the whole story. Thus, this concept of translucency denotes the blurriness characteristic of many practices in the policy world that are claimed to be transparent. Howell makes a similar point when she writes that 'reports of performed activities in actual areas or communities are embroidered to give the impression that meaningful communication and consultation have been taking place (e.g. UN-REDD Indonesia Semi-Annual Report 2011) when our observations show that this is often not the case' (2014: 265).

## GREEN CAPITALISM

As the discussion thus far shows, the practices of global climate mitigation efforts such as the UN REDD programme are grounded in the same market-liberal principles that have been behind the depletion of forests around the world in the first place. As Sullivan (2013) argues, the neoliberal reification of nature has led conservation into the realm of banking and finance, bringing conservation practices further into alignment conceptually, semiotically and materially with capital. With this logic, the conception of nature as capital, as an asset to be accounted for and accumulated, is increasingly normalised (Sullivan 2014).

Capitalism reframes nature as stock capital (Smith 1984; O'Connor 1994). According to Escobar:

> Through a new process of capitalization, effected primarily by a shift in representation, previously 'uncapitalized' aspects of nature and society become internal to capital ... capital thus develops a conservationist tendency, significantly different from its usual reckless, destructive form. (Escobar 1996: 326)

This rhetorical move from capitalism to green capitalism, of which UN REDD is an example, does not radically change the relationship between capital and forests or nature more generally. While capitalism defines nature as 'stock' for production and consumption, global environmental regimes define nature as 'stock' for carbon fixation. Attempts to reconstitute forests that have been decimated through capital accumulation are made through capital accumulation rationalities such as those embedded in UN REDD and CDM. Policy designers present the market as the best way to mediate and regulate environmental degradation. The logic is that with more market expansion, biodiversity conservation and development become compatible. However, experiences such as those we have discussed in this chapter demonstrate the fallacies of this logic.

The key concepts that facilitate this fusion of conservation with capital accumulation are sustainability and biodiversity. However, as Escobar remarks:

> the act of naming a new reality is never innocent. What views of the world does this naming shelter and propagate? Why has this new way of naming been invented at the end of a century that has seen untold levels of ecological destruction? (Escobar 1998: 55)

Sustainability and biodiversity are powerful historically produced discourses with a particular genealogy of knowledge. This genealogy is linked to the modernising discourse of neoliberal development (Sullivan 2006: 109; see also Argyrou 2005) – a view propagated by powerful international institutions such as the World Bank and the UN. Nazarea (2006: 318) rightly asks: 'is [biodiversity] a thing that exists in nature or just a conceptual and opportunistic sleight-of-hand to serve some hidden agenda? Is this thing that scientists and policy makers are preoccupied with recognized by local people?' The gap between Western ideas of forest conservation and local knowledge (Brosius 2000; West 2006; Escobar 2008) remains unresolved and the hegemonic discourse of conservation undisturbed.

Reframing nature in these green capital terminologies has benefited neither forests nor forest-dependent peoples; on the contrary, both are suffering from a triple burden: (1) extractive capitalism; (2) the consequences of climate change that has its sources elsewhere, mainly due to excessive material consumption in the rich, urban centres of the world; and (3) the market-based global environmental regimes established in the name of climate mitigation. As Sullivan (2013, 2014) remarks, the surplus value generated by the new ways of making 'nature' a commodity to be paid for will once again benefit those with the financial capital able to capture this surplus and maintain the historic dynamic of accumulation and inequality.

## CONCLUSION

In this chapter, we have outlined the consequences for forest peoples of using the principles of market environmentalism and capital accumulation to protect forest environments. We have highlighted how new trends that have emerged over the last few decades have led to the internalisation of earlier un-capitalised aspects of nature to the logic of capital. The new forest commodities such as carbon sinks release new financial value that can be traded, speculated on and profited from by those economically able to participate in this market. This is enabled by discourses of conservation, sustainability, biodiversity and climate change mitigation and by the silencing of alternative perceptions and past experiences of forest peoples with exclusionary and detrimental practices of nature management.

Delicate human-forest interactions that have been disturbed historically by colonialism, globalising capitalism and climate change are being further disturbed by global climate mitigation programmes. We have shown that UN REDD does not bring any substantial benefits to forest peoples, partly because the structure of global programmes intended to compensate for the environmental consequences of globalising capitalism draws on many of the same principles that have supported global economic expansion and extraction. On the one hand, embedded scientific management principles ignore the importance of local knowledge and negatively impact on local agricultural practices. On the other, as Howell points out, 'despite an increasing avowal to prioritise forest communities by national and international agencies involved, ... REDD pilot projects too often fail to be established in dialogue with local practices and aspirations: the local people are central and peripheral at the same time' (2015: 46).

The extension of global markets into forest areas, whether they commodify forest products intended for commercial exchange (timber, coal, pharmaceutical products, mineral ores) or commodify carbon sequestration in the name of environmental amelioration, redefine forest ecosystems as capital, ignoring their contribution to subsistence practices and local cosmologies. UN REDD is firmly situated within the matrix of global power 'where people become "human capital" and their habitat is quantified as "natural capital"' (Salleh 2010: 119). Therefore, the UN REDD programme runs the risk of increasing the exposure of local forest communities to global power structures and increasing pressures on both local livelihoods and local environments in the name of resolving a global environmental crisis – climate change – which is not of their making. UN REDD reproduces the errors of international economic development programmes that have 'classified populations, their knowledge, and cosmological life systems according to a Eurocentric standard ... [and have] legitimized relations of domination, superiority/inferiority' (Walsh 2010: 15).

It is no coincidence that the principle contributors to climate emissions are the same nations that are the proponents of the UN REDD programme. The pollution rights generated by UN REDD allow 'industrialized countries to delay long-term investment in no-carbon infrastructure' (Lohmann 2011) and inhibit actions aiming for radical changes in the economic system, the expansiveness of which has enveloped forests around the world. If progress is to be made on reducing the impacts of climate change, the rich OECD countries will have to move their own domestic aims from 'carbon neutrality' – a euphemism which misdirects responsibility for climate mitigation to people in the periphery – to rapid and deep carbon reduction in their own backyards.

## NOTES

1. In addition to the UN REDD programme, there are bilateral activities outside of the formal multilateral institutions. Norway, in addition to being the major donor of UN REDD, through its Climate and Forest Initiative (NICFI) aids REDD activities in countries such as Brazil, Indonesia, Tanzania, Guyana and recently Colombia.
2. In addition to Brazil.
3. The plus in REDD refers to four activities: (1) reducing emissions from deforestation; (2) reducing emissions from forest degradation; (3) sustainable management of forest; and (4) enhancement of forest carbon stocks.

## REFERENCES

Acosta, A. 2011. Extractivism and Neo-extractivism: Two Sides of the Same Curse. In M. Lan and D. Mokrani (eds), *Beyond Development: Alternative Visions from Latin America*. Amsterdam: Transnational Institute and Rosa Luxemburg Foundation, pp. 61–88.

Aguilar-Støen, M. 2017. Better Safe than Sorry? Indigenous Peoples, Carbon Cowboys and Governance of REDD in the Amazon. *Forum for Development Studies* 44(1): 91–108.

AIDESEP. 2010. Without Indigenous Territories, Rights and Consultation No REDD, Forests, Oil and Environmental Services Concession is Possible. *REDD-Monitor*. Available at: www.reddmonitor.org/2010/11/02/indigenous-peoples-organisation-in-peru-demands-an-indigenous-reddoutside-of-carbon-market-negotiations (accessed 1 September 2016).

Argawal, A. and S. Narain. 1985. *India: The State of the Environment*. Dehli: Centre for Science and Environment.

Argyrou, V. 2005. *The Logic of Environmetalism: Anthropology, Ecology and Postcoloniality*. New York and Oxford: Berghahn Books.

Borner, J., S. Wunder, S. Kanounnikoff, M.R. Tito and N. Nascimento. 2010. Direct Conservation Payments in the Brazilian Amazon: Scope and Equity Implications. *Ecological Economics* 69(6): 1272–82.

Brosius, J.P. 2000. Endangered Forest, Endangered People: Environemtalist Representations of Indigenous Knowledge. In R. Haenn and R. Wilk (eds), *A Reader in Ecology, Culture and Sustainable Living*. London: New York University Press, pp. 254–73.

Cabello, J. and T. Gilbertson. 2012. A Colonial Mechanism to Enclose Lands: A Critical Review of Two REDD+ Focused Special Issues. *Ephemera: Theory & Politics in Organization* 12(1/2): 162–80.

Carrere, R. and L. Lohmann. 1996. *Pulping the South: Industrial Tree Plantations and the World Paper Economy*. London: Zed Books.

Carus, F. 2010. British Deal to Preserve Liberia's Forests 'Could have Bankrupted' Nation. *Guardian*, 23 July.

Diemont, S.A., J.F. Martin, S.I. Levy-Tacher, R.B. Nigh, P.R. Lopez and J.D. Golicher. 2006. Lacandan Maya Forest Management: Restoration of Soil Fertility Using Native Tree Species. *Ecological Engineering* 28(3): 205–12.

Ellis, J.E. and D.M. Swift. 1988. Stability of African Pastoral Ecosystems: Alternate Paradigms and Implications for Development. *Journal of Range Management* 41(6): 450–9.

Escobar, A. 1996. Construction Nature: Elements for a Post-structuralist Political Ecology. *Futures* 28(4): 325–43.

—— 1998. Whose Knowledge, Whose Nature? Biodiversity, Conservation, and the Political Ecology of Social Movements. *Journal of Political Ecology* 5: 53–81.

—— 2008. *Territories of Difference: Place, Movements, Life, Redes, New Ecologies for the Twenty-first Century*. Durham, NC: Duke University Press.

Fernandes, W. and V. Paranjpye. 1997. Hundred Years of Displacement in India: Is the Rehabilitation Policy an Adequate Response? In W. Fernandes and V. Paranjpye (eds), *Rehabilitation Policy and Law in India: A Right to Livelihood*. New Delhi: Indian Social Institute, pp. 1–34.

Guha, R. 2006. *How Much Should a Person Consume? Environmentalism in India and the United States*. Berkeley, CA: University of California Press.

Hecht, S. and A. Cockburn. 2010. *The Fate of the Forest: Developers, Destroyers, and Defenders of the Amazon*. Chicago, IL and London: University of Chicago Press.

Hirsch, C., M. Aguilar-Støen and D. Mcneill. 2012. REDD+ and Indigenous Peoples. Insights from the Field. Policy brief, Centre for Development and the Environment, University of Oslo. Available at: www.eldis.org/organisation/A3497 (accessed March 2019).

Howell, S. 2014. No Rights – No REDD: Some Implications of a Turn Towards Co-benefits. *Forum for Development Studies* 41(2): 253–72.

—— 2015. Politics of Appearances: Some Reasons Why the UN-REDD Project in Central Sulawesi Failed to Unite the Various Stakeholders. *Asia Pacific Viewpoints* 56(1): 37–47.

Indian Government. 2006. *The Scheduled Tribes and Other Traditional Forest Dwellers (Recognition of Forest Rights) Act, 2006*. Delhi: Indian Parliament.

IPCC. 2014. *Climate Change 2014: Impacts, Adaptation, and Vulnerability.* Available at: www.ipcc.ch/report/ar5/wg2/ (accessed 20 September 2015).

Kalpavriksh, P.K. 2002. *The Forestry Encroachment Issue: A Briefing Note.* New Delhi: The Centre for Science and the Environment.

Kaus, A. 1993. Environmetal Perceptions and Social Relations in the Mapimí Biosphere Reserve. *Conservation Biology* 7(2): 398–406.

Kjosavik, D.K. and P. Vedeld. 2011. The Political Economy of Environment and Development in a Globalized World: An Introduction. In D.K. Kjosavik and P. Vedeld (eds), *The Political Economy of Environment and Development in a Globalized World.* Trondheim: Academic Press, pp. 17–43.

Klein, N. 2014. *This Changes Everything: Capitalism vs. the Climate.* New York: Simon & Schuster.

Lang, C. 2013. More Corruption Involving Norwegian REDD Funding in Tanzania? *REDD Monitor.* Available at www.redd-monitor.org/2013/02/06/more-corruption-involving-norwegian-redd-funding-in-tanzania (accessed 19 October 2015).

Lohmann, L. 2011. Capital and Climate Change. Review essay published on behalf of the Institute of Social Studies, The Hague. Available at http://onlinelibrary.wiley.com/doi/10.1111/j.1467-7660.2011.01700.x/ (accessed 15 November 2016).

Lynas, M. 2007. *Six Degrees: Our Future on a Hotter Planet.* London: Harper Collins.

Martinez-Reyes, J.E. 2004. Contested Place, Nature and Sustainability: A Critical Anthropo-geography of Biodiversity Conservation. The 'Zone Maya' of Quintana Roo, Mexico, Anthropology. Graduate School of the University of Massachussets.

Mathews, A. 2014. Imagining Forest Futures and Climate Change: The Mexican State as Insurance Broker and Storyteller. In J. Barnes and M. Dove (eds), *Climate Cultures: Anthropological Perspectives on Climate Change.* New Haven, CT: Yale University Press, pp. 199–220.

Murombedzi, J. 2003. Devolving the Expropriation of Nature: The 'Devolution' of WildlifeManagment in Southern Africa. In W. Adams and M. Mulligan (eds), *Decolinising Nature: Strategies for Conservation in a Post-colonial era.* London: Earthscan, pp. 135–51.

Nazarea, V. 2006. Local Knowledge and Memory in Biodiversity Conservation. *Anjual Review Anthropology* 35(1): 317–35.

Newmann, R.P. 1998. *Imposing Wilderness: Struggles over Livelihood and Nature Preservation in Africa.* Berkeley, CA: University of California Press.

Nustad, K.G. 2011. Performing Natures and Land in the iSimangaliso Wetland Park, South Africa. *Ethnos*76 (1): 88–108.

O'Connor, M. 1994. *Is Capitalism Sustainable?: Political Economy and the Politics of Ecology, Democracy and Ecology.* New York: Guilford Press.

Okamoto, T., C. Hirsch and M. Aguilar-Støen. 2012. REDD+, Indigenous Peoples, the Role of the State, NGO and Other Actors. Monitoring,

Implementation and Tenure Rights. Insights from the Field. Policy brief, Centre for Development and the Environment, University of Oslo. Available at: www.eldis.org/organisation/A3497 (accessed March 2019).

Olding, W. 2017. *Norway's International Climate and Forest Initiative: Lessons Learned and Recommendations*. Oslo: NORAD.

Oosthoek, K. 2007. The Colonial Origin of Scientific Forestry in Britain. Environmental History Resources, 25 June. Available at: www.eh-resources.org/colonial-origins-scientific-forestry (accessed 15 November 2016).

Osborne, T. 2011. Carbon Forestry and Agrarian Change: Access and Land Control in a Mexican Forest. *Journal of Peasant Studies* 38(4): 859–83.

Salinas, C.G. 2011. *Añorada Esperanza: Respuestas Locales a las políticas neoliberales. Uruguay y la industria de la celulosa*. Buenos Aires: Editorial Antropofagia.

—— 2016. Elusive Appearances: How Policies Fail in the Argentinian Paraná Atlantic Forest. PhD thesis, Department of Social Anthropology, University of Oslo.

Salleh, A. 2010. Climate Strategy: Making the Choice Between Ecological Modernisation or Living Well. *Journal of Australian Political Economy* 66: 118–43.

Schiavoni, G. and A. Alberti. 2014. Autonomy and Migration: The Forestal Workers of the Northeast of Misiones (Argentina). *Sociología del trabajo* 23(Winter), 169–77.

Shearman, P., J. Bryan, J. Ash, B. Mackey and B. Lokes. 2010. Deforestation and Degradation in Papua New Guinea: A Response to Filer and Colleagues, 2009. *Annals of Forest Science* 67(3): 300–4.

Smith, N. 1984. *Uneven Development: Nature, Capital, and the Production of Space*. Oxford: Blackwell.

Sullivan, S. 2006. The Elephant in the Room? Problematising 'New' (Neoliberal) Biodiversity Conservation. *Forum for Development Studies* 33(1): 105–35.

—— 2013. Banking Nature? The Spectacular Financialisation of Environmental Conservation. *Antipode* 45(1):198–217.

—— 2014. The Natural Capital Myth; or Will Accounting Save the World? Preliminary Thoughts on Nature, Finance and Values. The Leverhulme Centre for the Study of Value, Working Paper series 3, 2–36.

Sunderlin, W.D., C.D. Pratama, A.B. Bos, V. Avitabile, E.O. Sills, C. de Sassi, S. Joseph, M. Agustavia, U.A. Pribadi and A. Anandadas. 2014. *REDD+ on the Ground: A Case Book of Subnational Initiatives Across the Globe*. Indonesia: Center for International Forestry Research.

Tsing, A.L. 2005. *Friction: An Ethnography of Global Connection*. Princeton, NJ: Princeton University Press.

Walsh, C. 2010. Development as *Buen Vivir*: Institutional Arrangements and (De)colonial Entanglements. *Development* 53: 15–21.

WCED. 1987. *Our Common Future*. Oxford and New York: Oxford University Press.

West, P. 2006. *Conservation is Our Government Now: The Politics of Ecology in Papua New Guinea*. Durham, NC: Duke University Press.

Wilhite, H. 2014. Les Effets de l'étau du Développement et des Pressions Environnementales sur le Mode de Vie des Populations des Forêts Indiennes. In C. Bréda, M. Chaplier, J. Hermesse and E. Piccoli (eds), *Terres (Dés) humanisée: Ressourses et Climats*. Louvain-la-Neuve: Academia-L'Harmattan s.a.

—— 2016. *The Political Economy of Low Carbon Transformation: Breaking the Habits of Capitalism*. London: Routledge.

# 10. Climate Change, Oceanic Sovereignties and Maritime Economies in the Pacific

*Edvard Hviding*

CLIMATE CHANGE AND DIMENSIONS OF
THE UNKNOWN IN THE PACIFIC OCEAN

For the world at large, diverse predictions about global climate change point in one direction: rising sea levels, flooding, drought and recurrent extreme weather will make coastal areas gradually uninhabitable, leading to the submersion of urban centres, migrations of displaced people, and economic and political upheaval (Nicholls and Cazenave 2010; Vidas 2011). Global climate crisis scenarios are habitually seen as future concerns, but this is not the case for the islanders of coral atolls in the Pacific (Barnett and Adger 2003; Lazrus 2012): they are in danger now, and the islands and islanders of the Pacific are on the world's climate change frontline. They contribute the least to global warming, but are set to suffer the most from its effects, and now undergo challenging local trial runs for a global future severely impacted by global warming. For Pacific Islanders, then, future climate crisis is already there – *the future is today*.

This chapter focuses on the precarious future of island nations in terms of the extraordinary maritime scale of Pacific states, through which their spatial sovereignty and financial viability are grounded primarily in their Exclusive Economic Zones (EEZs), vast ocean expanses radiating 200 nautical miles from the outermost inhabited land, however low-lying it may be, and legally defined by the United Nations Convention of the Law of the Sea (UNCLOS; UN 1982). In recent fieldwork on the stages of international diplomacy, I have frequently heard comments by Pacific Islands state leaders and diplomats to the effect that the UNCLOS is the world's greatest gift to the Pacific – the Convention in fact secures the economies of the 'small island developing states' (SIDS) of the region, and allows them to self-identify as 'Big Ocean States'.

The Pacific Ocean covers about a third of the world's surface but is home to only about 10–11 million people. However, the people and islands of Oceania, as the islanders themselves prefer to call their region (Hau'ofa 1994), represent extraordinary cultural and biological diversity (Lal and Fortune 2000). The great ocean which islanders see as the foundation of their unique existence as 'people of the sea' sustains some of the world's most valuable marine resources, mainly migratory tuna (D'Arcy 2006; Bell et al. 2011). Their islands are often small and low-lying but are important legal baselines of EEZs, usually disproportionate to the tiny land mass. In line with this land-sea equation, island nations often express sovereignty with particular reference to their ocean – reflecting a maritime connection typical of Pacific world views (Hau'ofa 1994; Hviding 2003, 2015) and suggesting, in the words of former Kiribati president Anote Tong (2015), the view of small island states as big (or large) ocean states.

To sharpen the analytical approach I shall invoke, and later discuss in some detail, the concept of *mare nullius*, 'ocean owned by no one' – a novel extension of the *terra nullius* of colonial times, which underwrote the imperial dispossession of land (for example, Fitzmaurice 2007) in Australia and other places. *Mare nullius* has been described for the settler states of Canada and Australia as the 'colonial negation of indigenous tenure and authority via the doctrine of *terra nullius*', so that '[t]he *mare nullius* of seas ... declared vacant of indigenous tenure or authority by colonisers, became subject to European representations of seas as international commons and ... the incorporation of offshore areas as the sovereign jurisdiction of states ... ' (Mulrennan and Scott 2000: 682). In a major new research project, I work with a multidisciplinary team to take this so far undeveloped concept as our core point of orientation for empirical investigation, interdisciplinary analysis and diplomatic engagement, expanding *mare nullius* to an appropriate associative icon for the distinctive Pacific entanglements of sea level rise and sovereignty.[1]

The ocean is indeed fundamental to the historical and contemporary development of local and regional cultures, identities and ways of life in the Pacific Islands (D'Arcy 2006; Matsuda 2012; Hviding 2015). In the twenty-first century, however, indigenous Pacific views of the great ocean as generative and supportive of regionally specific ways of human existence are fundamentally challenged as the tables are turned: for the islanders of the great Pacific, the rapidly encroaching effects of global climate change are transforming the life-giving ocean into a threat. As the warming, acidification and rising of the sea erode coral reefs and coastal zones throughout the region, and as new patterns of extreme weather become regular, low-lying atoll nations of the central Pacific seem destined for an unprecedented political situation with the permanent, partial or total flooding of their land. From interpretations of the UNCLOS, it is unclear whether diminishing or disappearing national territories will imply a similar fate for the EEZs of Pacific atoll nations (consider the EEZ of Kiribati at 3.5 million km$^2$, contrasted to a total land area of 810 km$^2$). However, if a contraction of land should imply a concomitant contraction of EEZs, displaced atoll populations stand also to lose their primary economic resource for the future in global terms. This predicament is based on the fact that more than 60 per cent of the world's tuna swim in regular migratory fashion through the EEZs of the atoll nations of the central Pacific.

A new set of challenges and contestations are emerging, then, relating to future state and maritime sovereignties at indigenous, national and regional levels – and to the potential for predatory initiatives into potentially sovereign-less ocean from distant-water fishing nations and the global capitalist connections of high seas tuna fisheries. New initiatives in the law of sovereignty and the UNCLOS may be expected, associated with appeals to climate justice from the Pacific Islands on the global scenes of climate change negotiations.

I have carried out altogether about three and a half years of fieldwork in the Melanesian archipelago of Solomon Islands – where in fact I started out with matters of the sea, studying how in the largest coral lagoon in the world – the Marovo Lagoon – every square metre of submerged reef, low coral islands and tidal zones were subject to complex forms of customary tenure (Hviding 1996, 1998), counter to ideas still prevailing in the 1980s that land, not sea, could be owned. That research took place in the context of a vibrant maritime-based subsistence economy. In recent years, however, I have had the opportunity to work on what may be termed a pan-Pacific scope, seeking to develop elements of a truly Oceanic anthropology of the present time (Hviding 2016). I have been spending days and weeks in the capitals and offices of government and diplomacy of Fiji, Samoa, Cook Islands and other Pacific nations, devising impact-oriented performances with Pacific artists and activists, and talking to Pacific diplomats and climate change negotiators in New York, Brussels and Paris; in effect, doing short but intensive fieldwork in a myriad of urban, bureaucratic and activist contexts where political affairs are discussed and negotiated and policies decided and implemented.[2] From a solid initial grounding in one Pacific locality, it is possible, I believe, to work truly regionally, in so far as just about all politics of some scale in the Pacific are international and mediated through regional organisations like the Pacific Islands Forum, and in so far as cultural diversity across the entire region may, from a certain irreverent perspective (shared by many of the Pacific's indigenous frequent travellers), be seen as variations on central themes. This, then, is the anthropological position from which I try to grapple with some of the most pressing issues of contemporary Oceania. And that work has led me back to the sea again, but now into the large-scale maritime economies of the contemporary Pacific, where nearly two-thirds of the world's tuna resources are managed through regional organisations that provide substantial proportions of the annual national budgets of many island nations.[3]

Sea level rise, which in some parts of the Pacific Ocean accelerates at a rate of up to three times that of the world average, directly threatens those Pacific nations – the Marshall Islands, Kiribati, Tuvalu and the New Zealand-governed territory of Tokelau – whose land is composed entirely of low-lying atolls. If predictions are taken to the extreme, the partial or total disappearance of the lands on which state formations are based will pose unprecedented threats to the continued sovereignty of those states. To add to the complexity of such issues for the Pacific region in general, the larger nations, whose land mainly consists of high islands, also experience threats to the continued definitions of EEZs, as these are often based on outlying low islands. This variety of predicted but unprecedented transformation of the basis for sovereign seas poses severe challenges to prevailing international law. I have already noted that this body of law at the outset is far from clear on these emerging issues. The United Nations Convention of the Law of the Sea (UN 1982) is, for example, not ratified by the United States, and the Convention's legal implications are in some cases poorly defined in terms of international maritime boundaries between island nations. The future role of the great Pacific Ocean in a radically transformed island environment is thus, so far, a largely unknown dimension.

## LIVES AND ECONOMIES IN AN OCEANIC REGION

I now provide a regional anthropological sketch of Oceania for readers not so familiar with the region, and I include a glimpse of the recent and present-day growth of what are globally distinctive forms of maritime economies based on tuna fisheries.

In absolute terms, the Pacific Islands provide some globally unique examples of the interactions between resilient subsistence economies, high marine and terrestrial biodiversity in environments largely under customary communal ownership, and threats posed by the long-term but already significant local effects of climate change, other forms of more rapid environmental transformation, and any form of natural disaster, predictable or not. Oceania is the most dispersed, and biologically and culturally, the most diverse region of the world. Across thousands of kilometres in the equatorial and southern Pacific Ocean, small and large groups of islands constitute culturally distinct nations with populations ranging from a few thousand (for example, Tuvalu, Tokelau, Niue) to several million (Papua New Guinea). Subsistence-based rural economies based on small-scale fishing and agriculture (and the resources of coral reefs and rainforests) remain strong and vital in many Pacific nations, perhaps to a degree unsurpassed elsewhere in the world, since up to 85 per cent of national populations live largely subsistence-based village lives. However, rapid inroads are being made by global agents of large-scale resource extraction. Pacific fish, timber and minerals are subject to such global desire, and only to a small degree are Pacific nations and their people adequately compensated for the resources extracted (D'Arcy 2013).

The only real exception is a remarkably efficient and quite profitable regional, home-grown but scientifically managed organisation for the regulation of industrial tuna fisheries – the so-called Parties to the Nauru Agreement (PNA), emblematic of Oceania's regional maritime economy and collectively managing access to the EEZs of the Federated States of Micronesia, Kiribati, the Marshall Islands, Nauru, Palau, Papua New Guinea, Solomon Islands and Tuvalu – the eight largest ocean states of the Pacific. More specifically, PNA controls, for example, 50 per cent of the world's total annual catch of skipjack tuna for canning.[4] The PNA sets regulations on allowable fishing technology, and under its 'Vessel Day Scheme' (VDS) arrangement, auctions off to overseas fishing operators a predetermined number of annual vessel fishing days. For 2015, the average set fee per vessel day was US$8,000, and the set number of vessel days was approximately 45,000. Actual revenue from auctions often exceeds set fees, and in recent years, vessel days have been sold for up to US$15,000. From an annual catch in its member countries' EEZs worth up to US$3,000 million, the PNA for 2016 was reported to have generated a revenue of about US$540 million – a percentage of revenue from total catch that through aggressive and successful PNA politics increased from about 3 per cent in 2010 to 18 per cent in 2016. The annual income from the VDS system is distributed to the member nations on the basis of EEZs and resource stock estimates. For example, the small nation of Kiribati receives more than 20 per cent of total PNA income from fees (Clark and Clark 2014). Patterns in recent years have remained consistent with these figures.

The success of the PNA notwithstanding, inroads by the global political economy generate an increasing range of environmental problems in Oceania, and urban sectors are rapidly growing, without any infrastructure development necessarily following. For example, in the remarkable but precarious atoll cities of Tarawa and Majuro, capitals of Kiribati and the Marshall Islands, half of the populations of those two nations (about 50,000 in Tarawa and 30,000 in Majuro) lead vulnerable urban lives with severe challenges to the everyday provision of electricity and fresh water, on slivers of low coral land. It is clear that the Pacific Islands region is considered particularly vulnerable to climate change also because of its special geographical features, its inhabitants' dependence on their immediate natural environments and its lack of technological buffers.

Let us now look more closely at the region in terms of three very different interpretations of the great maritime continent of Oceania:

1. The anthropological definition, derived from the days of European exploratory voyages, of the three culturally distinct sub-regions of Melanesia, Polynesia and Micronesia – each with its set of assumed characteristics that somehow obscure the widely shared patterns of long-term cultural history and the current forms of pan-Pacific regionalism (Figure 10.1).

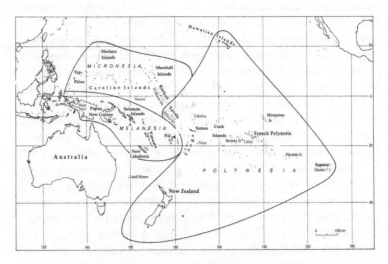

*Figure 10.1* The anthropological boundary-making of Oceania

Source: University of Bergen.

2. The division of the ocean into EEZ grounded in the far from uncontested 200-mile jurisdiction defined by the UNCLOS, providing vast sovereign territories to nations with negligible land areas (Figure 10.2).

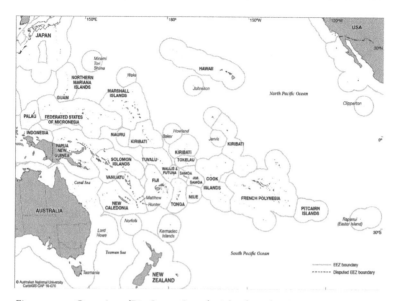

*Figure 10.2*   Oceania as 'Big Ocean States' – islands and EEZs

Source: CartoGIS Services, College of Asia and the Pacific, the Australian National University.

3. The unbounded 'Sea of Islands' proposed by prominent indigenous thinker and Tongan anthropologist, the late Epeli Hau'ofa (1994) – a strong alternative Oceanic vision emphasising pan-Pacific interconnectedness and travel rather than isolation and remoteness. Basically, a third of the globe and invariably, in terms of conventional cartography, its backside (Figure 10.3).

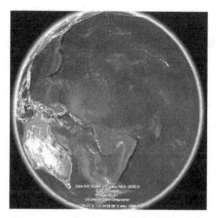

*Figure 10.3*   Oceania as a 'Sea of Islands'

Source: Google Earth.

From the first stages of colonial history, the peoples of the far-flung Pacific Islands had a long and distinguished record of meeting the peoples of other regions of the world on Pacific social and cultural terms (for example, Thomas 2010; Igler 2013). Some 'first contact' narratives from the region are both entertaining and sobering. When the British ship *Alexander* passed by the small island of Simbo in Solomon Islands on 8 August 1788, the islanders, who could not possibly have encountered a European ship before, came out in canoes and engaged in immediate on-board trade – local shell artefacts for iron objects. They were not amazed, nor were they armed – they were reported to be 'without any visible apprehension' (Dureau 2001: 133 quoting the log of the British convict transport ship *Alexander*).

Pacific Islanders continue to engage with outsiders through their own Pacific economic and political positions, priorities and social (and diplomatic) protocols. The ocean has channelled and mediated much of this in the great long run of history. But as already noted, while the Pacific is the world region least responsible for global warming, it is the one first affected by the consequences of environmental and climate crises. Droughts, floods, coral bleaching, shoreline erosion, seawater inundation and unpredictable extreme weather are only some of the challenges faced today by islanders, and the local interpretations of these globally grounded processes of irreversible change are as diverse as the region itself (for example, Rudiak-Gould 2013).

In fact, the effects of climate change in Oceania are only likely to exacerbate the challenges Pacific Islanders have come to know and deal with over thousands of years, posed by volatile environmental forces which make volcanic eruptions, earthquakes, tsunamis and tropical storms potentially devastating but quite regularly occurring phenomena. The islanders of the Pacific, well used as they are to dealing with so many destructive environmental forces, are far from unprepared to tackle the onset of climate change. As documented in detail and diversity in physical geographer Patrick Nunn's remarkable book *Vanished Islands and Hidden Continents of the Pacific* (2009), in which he combines geology and mythology, the islanders of the great ocean have lived with a high level of environmental instability since they first arrived from the archipelagoes of southeast Asia and settled thousands of years ago, and the Pacific's deeper time of geological scale has seen extraordinary transformations of sea and land.

## CHANGES IN THE WEATHER: THE FUTURE IS TODAY

In 1989, as global messages about the greenhouse effect were just beginning to reach the local level in places like the Solomon Islands in the tropical southwestern Pacific, an inquisitive, well-educated and voraciously well-read school teacher in the Marovo Lagoon – the place where I have been doing regular fieldwork since 1986 – expressed his concerns over changes in weather patterns. He explained to me his own views of continuity and change in the weather, as well as his attempts to make sense of these processes as they were visible on a daily basis from the seaside veranda of his family's house, built of sago palm leaves and forest materials:

> The people of old told us how things are. They said that this wind will come at that time and finish after so many moons, and then the other wind will take over. All other things too, tides, rain, and all the fish in the sea, and whatever, follow this, they said. And so, all things have their time, and because the people of old knew how to mark those times, they told us about this, and we believed them, because we could see this with our eyes, too. But now, I don't know. It seems that the weather is not straight any more. I still would like to trust what the people of old taught us, but one day I came to think that maybe they fooled us back then? Maybe things have never been as straight as they would want us to believe! (Hviding 1996: 374)

That was the final paragraph of my first published monograph. Now, almost 30 years after those words were spoken, sea level rise, droughts and recurrent extreme weather threaten to make coastal areas and ecologically sensitive lands across the world uninhabitable, leading to large-scale migrations of displaced people, flooding of the globe's major urban centres and resultant economic and political upheavals. Future in the Anthropocene is uncertain, to say the least, and as discussed in detail in this book as a whole, those uncertainties will likely provide extraordinary opportunities for capitalism without restraint. Or is the future uncertain? For whom?

I noted in the introductory pages to this chapter that for the islanders of low-lying coral atolls in the Pacific, there *is* no uncertain future of local upheavals caused by global climate change. For them, that future is there today. The tropical western-central Pacific experiences current rates of sea level rise up to three times the world average – a result of the combined forces of changing wind patterns, with persistent trade winds blowing more strongly than before from the southeast, unprecedented warming of the ocean, and other factors that together drive massive volumes of water to pile up along a southeast to northwest axis. During a conversation, an oceanographer friend and colleague drew a cross-section of the central Pacific for me with a substantial slope rising up from east towards west. The ocean is not flat.

Some regularities still persist in Pacific wind patterns. Every year in the period from December to March, the southeasterly trade winds are replaced by strong northwesterly winds – often described as the Pacific tropical monsoons – blowing with full force and bringing those huge volumes of piled-up ocean with them towards the low-lying archipelagoes of central tropical Oceania. While these monsoons were always part of the annual cycle in the tropical Pacific, now they are stronger and bring far more ocean with them. Already, the ocean has through repeated floods obliterated arable soil and freshwater sources and undermined continued human existence, such as in parts of the Marshall Islands and Kiribati. The processes discussed here converged in a particularly acute and concrete way in the central Pacific on 3 March 2014, when Majuro, the capital of the Republic of the Marshall Islands, was brutally jarred awake by a rising sea. That morning an extreme flood, locally referred to as a 'king tide', caused storm waves to pass over the reef and smash with full force into the oceanside neighbourhoods of this low-lying, improbable urban configuration of 30,000 inhabitants, on a land mass of only a few square kilometres, where elevation is nowhere more than 2.5 m. As waves hit the shore, their impact dug away at the ground. The foundations of concrete houses were left hanging in thin air, later to collapse. The bones of ancestors were exposed as waves broke their tombs apart. The homes of hundreds of people were destroyed, and a national state of emergency was declared.

Pacific climate researchers have predicted that high-tide floods and storm surges, such as the one of 3 March that built up near Japan and hit the atolls of the central Pacific, will only increase in frequency and magnitude in the coming years. The pattern has indeed continued in the low-lying nations of the Marshall Islands, Kiribati and Tuvalu, with storm floods from the northwest now occurring several times every year. Anthropologist Ingrid Ahlgren, who has lived in the Marshall Islands for much of her life, has reported how proverbs remind Marshall Islanders that

> their surrounding ocean could present the unexpected, without warning or care for the land dwellers. [That] meaning is co-mingled with warfare – as many Marshallese proverbs do – here, the environment is similar to the way an enemy can arrive from the ocean without warning. (Ahlgren 2014)

Truly, the life-giving ocean of Oceania's history can turn on its inhabitants and intervene terribly in their lives. Islanders and their governments try to respond to this dire predicament of a future that is already there. As the atoll states are now planning purchases of land on larger, mountainous islands in other Pacific nations (Kiribati bought 20 km$^2$ of land in Fiji in 2014), as EEZs once defined forever by the UNCLOS become unstable from the shrinking area of dry land – and as the uncertain status of 'climate change refugee' is being contested in international and indigenous contexts – new identities, sovereignties, economies and socialities are already apparent on the Pacific scene. The islanders of the Pacific are not only the 'poster faces' of global climate change peering at the world from the covers of reports and social media feeds; they are also undergoing today the challenging local trial runs for a global future already impacted by the effects of global warming. Pacific Islanders are challenged to find ways of living without their ancestral lands, in new localities where they will be migrants and perhaps ending up in an unprecedented situation where, as expressed by Marshall Islands poet Kathy Jetnil-Kijiñer (2017: 70):

… you, your daughter
and your granddaughter too
will wander
rootless
with only
a passport to call home

Interestingly, for parts of Oceania that future of today is also one of the past, as attested to by colonial experiments of relocation and resettlement of islanders, long before climate change displacement became a concern in public and political forums (for example, Teaiwa 2015; Tabe 2016). Today, this past of colonial violence in the central Pacific, involving dispossession, evictions, secretive operations of nuclear testing and forced migrations with no return and revoked citizenship, fuel dystopian and utopian images for a future already here, as rising seas wash away ancestral bones, undermine the basic needs of human habitation and threaten sovereignty.

## *MARE NULLIUS* – NOBODY'S OCEAN?

With the low lands of Oceania facing such an immediate threat of disappearing under the sea, with the coasts of higher islands eroding, and with few technological countermeasures (for example, of the Dutch type) at hand, the questions I posed in my introduction take centre stage. What happens with the 200-mile EEZ of a nation whose lands, from which the extent of those 200 nautical miles is measured, are disappearing? The question is first on record as having been raised in 2009 by Phillip Muller, then the Ambassador of the Marshall Islands to the United Nations (and more lately Foreign Affairs Minister). His scenario, acutely observed from the role of his nation as a canary bird of global warming and sea level rise – and indicative of the 'future is today' stance I have referred to – was this:

> the seas are rising, and some decade – no one knows which – his country of twenty-nine coral atolls and five islands, located midway between Hawaii and Australia, is going to be under water. When that happens, a number of novel legal questions will arise. If a country is under water, is it still a state? Does it still have a seat at the United Nations? What becomes of its exclusive economic zone, and the fishing rights on which it depends for much of its livelihood? What countries will take in its displaced people, and what rights will they have when they arrive? Do they have any recourse against those states whose greenhouse gas emissions caused this plight? (Gerrard and Wannier 2013: xvii)

Ambassador Muller presented this complex scenario, eloquently evoking nearly every aspect of present-day climate change negotiations from 'loss and damage' to the situations of 'climate change refugees' and any issue in between, to Columbia University's new Center for Climate Change Law. The response was a conference and a weighty 639-page tome of legal analysis (Gerrard and Wannier 2013) published by Cambridge University Press, and leaving readers in no doubt that legal scholars are prone to writing very detailed papers of 40 or more densely footnoted pages. While the book is a benchmark publication covering almost every conceivable legal issue concerning territorial sovereignty, human displacement and resettlement in the Pacific, and what is now referred to in climate change negotiations as loss and damage, it is remarkable that there is no reference to fisheries, to Oceania's maritime economies, nor even once to the single most valuable marine resource of the Pacific Ocean – the 60 per cent of the world's tuna swimming there and migrating through the threatened EEZs.[5]

From years of observation of the Pacific regional fisheries scene, partly facilitated by personal acquaintances with some of its most powerful players, it is clear to me that the reason Ambassador Muller and so many other Pacific Islands leaders are nervous about the future of their EEZs is precisely that there are so many foreign forces just waiting for the opening-up of the tuna fishing grounds to the world at large. For decades, the so-called 'high seas pockets' where EEZs do not overlap have been contested scenes to which distant-water fishing fleets – including an increasing number of surplus Spanish purse-seining vessels operating on behalf of the European Union – seek and often obtain access, thereby harvesting the tuna passing through on the way from one Pacific EEZ into another. This practice has been subject to heated conflict, particularly with the Pacific regional fisheries management organisations, including the aforementioned PNA, whose management mechanisms designate the high seas pockets as no-fishing conservation zones.

To analyse this situation of predatory capitalism, some would say neocolonialism, in the ocean, it is instructive to look at a terrestrial counterpart. We know from the history of the Pacific that the Roman law concept of *terra nullius* or 'nobody's land' – land that is unclaimed, unowned, uninhabited and so on – played a large part in the insidious and destructive colonisation of parts of Oceania. Captain Cook had no doubt that he found *terra nullius* when first landing in Australia in 1770. No real people lived there, as he and his imperial contemporaries saw it. As for the ocean, the Dutch legal philosopher Hugo Grotius had coined the concept of *mare liberum* already in 1609 – implying that because of the needs of Europe to send its ships around the world to carry out trade and imperialism, the sea really had to be free. According to Grotius, the sea could not be owned, the sea was not claimed by anyone and therefore European vessels – in particular Dutch vessels, it should be said – were seen to be entitled to travel anywhere, ply their trade, fire their guns and carry out their predations on behalf of imperial conquest.

In Oceania, at least in the 12 independent island nations (and the two self-governing ones associated with New Zealand), the sovereign state is universally a creation of the indigenous peoples themselves, and so *mare nullius* – 'nobody's ocean' – is also an apt designation for those parts of the Pacific Ocean *not* under any sovereign maritime jurisdiction, that is, outside of and in between EEZs, known today as 'Areas Beyond National Jurisdiction' and currently subject to intense negotiations at the United Nations. If EEZs, then, diminish when lands contract as a result of sea level rise, so *mare nullius* increases, and industrial fishing fleets from China, Japan, Korea, Indonesia, the United States and distant Europe (to mention the major suspects) will be ready to expand their activities without interference from what is frequently seen as the pesky and irritating PNA. It is safe to say that the tuna-hungry non-Pacific parts of the world have every interest in seeing the PNA and the EEZs it controls diminished.

There is another incentive for promoting *mare nullius* definitions of the Pacific Ocean: the increased attention given to seabed mining of manganese nodules and other mineral resources. At regional meetings in the Pacific, it is often remarked of the French so-called 'Overseas Country and Territory' (OCT) of French Polynesia that if only the strong vested interests in the tuna of the OCT's almost 5 million km$^2$ of EEZ could be abandoned as an economic and political factor, there would be direct access to deep sea mining.

The EEZs produced by the ratification of the United Nations Convention of the Law of the Sea have generated sovereign territories in the Pacific Ocean. Those huge sovereign territories, so crudely outlined as 200-mile zones, are measured from the furthest island in any part of an archipelago, including very remote and territorially profitable outlying islands that may not even be populated. And thus the nation of Kiribati has several huge patches of EEZ totalling almost 3.5 million km$^2$ of ocean; and the Republic of the Marshall Islands has just about 2 million km$^2$. These atoll nations with their small populations, and globally negligible areas of land so vastly important for their people, in fact have some of the largest EEZs in the world. Kiribati ranks as number 12 on the global list of EEZs, the Federated States of Micronesia is number 14, and the Marshall Islands ranks as number 20. It goes without saying that population-to-EEZ ratios are staggering – and for a nation like Kiribati, with a population of 115,000, it is the EEZ that provides the major source of state revenue, through the Vessel Day Scheme of the PNA, which has such capacity to irritate distant-water capitalist fishing powers.

## CONCLUSION: THE OCEAN OF THE ISLANDERS

It is clear that the Pacific Ocean has never been *mare nullius* for its inhabitants. That ocean was never 'nobody's ocean'. Through past and present travel and connections on inter-island and inter-archipelago scales, even the high seas outside of legally defined EEZs are named and known. In his comparative study of maritime cosmologies and histories in Oceania, Paul D'Arcy (2006: 152) comments how the regular practising of wide-ranging inter-island ties in the far-flung coral atolls of the Caroline Islands caused the ocean to be 'conceived of as named sea lanes between specific destinations and named open sea outside sea lanes', and Ingrid Ahlgren (2014) has reported for the twin atoll chains of the Marshall Islands that the apparently open ocean beyond sight of land is criss-crossed with named, known and precisely located seamounts (sunken islands) that guide navigation in what to outsiders seems a deep dark void. The UNCLOS has in a sense codified these extreme horizons of human maritime knowledge into even wider legal realms of sovereignty with huge economic potential for the 'Big Ocean States'. It is this potential that is likely to be contested when sea levels rise and challenge sovereignties that are from the outset far from exact and uncontested.

## NOTES

1. The research project *Mare Nullius? Sea Level Rise and Maritime Sovereignties in the Pacific – An Expanded Anthropology of Climate Change* involves anthropology, law, climate science, political science and other fields, and is funded by the Research Council of Norway for 2018–23 (Grant No. 275312). See the project website at www.uib.no/en/marenullius (accessed 5 February 2019).
2. I was given these opportunities for a new style of anthropological fieldwork in my capacity during 2012–16 as coordinator of the EU-funded European Consortium for Pacific Studies (ECOPAS, www.pacific-studies.net, accessed 5 February 2019), a collective of four European and two Pacific research centres focused on building Europe-Pacific dialogue particularly concerning the human dimensions of climate change (Hviding 2016). Since 2017 I have increasingly participated as an accredited observer at regional intergovernmental meetings in the Pacific and global negotiations at the United Nations in New York, thanks to the generosity of Pacific Islands diplomatic missions and to requests from Norway's Ministry of Foreign Affairs to provide research-based advice on a multitude of Pacific issues.
3. As I proceed with analysis of these large-scale regional maritime economies, the comparative paucity of references to published literature is evident. This is a consequence of a scarcity of relevant published literature on what are very recent developments for the Pacific region. Debates and reporting still take place mainly in digital and printed media and not in the academic literature, and to some degree in online 'Policy Brief' reports from regional Pacific and United Nations organisations.
4. Notably, the PNA's management system is also globally certified by the Marine Stewardship Council as the world's largest sustainable tuna purse seine fishery. See the PNA website at www.pnatuna.com (accessed 5 February 2019).
5. Although scientific predictions of significant warmer ocean waters indicate that the different species of Pacific tuna may change their annual migration patterns, and move in more easterly waters, outside of the EEZs of the central tropical Pacific. Thus, global warming and climate change may deprive those island nations of important tuna stocks simply through the response to warmer seas by the fish themselves (Bell et al. 2011: 435–80).

## REFERENCES

Ahlgren, I. 2014. Perilous Rhetoric? Perceptions and Histories of Environmental Threats in the Marshall Islands. Paper presented to session on *Small Islands in Peril or Under Pressure*, Annual Meetings of the Association for Social Anthropology in Oceania, Kona, Hawai'i.

Barnett, J. and W.N. Adger. 2003. Climate Dangers and Atoll Countries. *Climatic Change* 61: 321–37.

Bell, J.D., J.E. Johnson and A.J. Hobday (eds). 2011. *Vulnerability of Tropical Pacific Fisheries and Aquaculture to Climate Change*. Noumea, New Caledonia: Secretariat of the Pacific Community.

Clark, L. and S. Clark. 2014. The PNA Vessel Day Scheme. A presentation to AustralianNational University Pacific Update, Canberra, 16–17 June.

D'Arcy, P. 2006. *The People of the Sea: Environment, Identity, and History in Oceania*. Honolulu: University of Hawai'i Press.

—— 2013. The Nourishing Sea: Partnered Guardianship of Fishery and Seabed Mineral Resources for Economic Viability of Small Island Nations. *Sustainability* 5: 3346–67.

Dureau, C.M. 2001. Recounting and Remembering First Contact on Simbo, Western Solomon Islands. In J.M. Mageo (ed.), *Cultural Memory: Reconfiguring History and Identity in the Postcolonial Pacific*. Honolulu: University of Hawai'i Press, pp. 130–62.

Fitzmaurice, A. 2007. The Genealogy of Terra Nullius. *Australian Historical Studies* 38: 1–15.

Gerrard, M.B. and G.E. Wannier (eds). 2013. *Threatened Island Nations: Legal Implications of Rising Seas and a Changing Climate*. Cambridge: Cambridge University Press.

Grotius, H. 1609. *Mare Liberum*. Original in Latin; published in English in 1916 as *The Freedom of the Seas*. New York: Oxford University Press.

Hau'ofa, E. 1994. Our Sea of Islands. *The Contemporary Pacific* 6: 148–61.

Hviding, E. 1996. *Guardians of Marovo Lagoon: Practice, Place, and Politics in Maritime Melanesia*. Pacific Islands Monograph Series, Vol. 14. Honolulu: University of Hawai'i Press.

—— 1998. Contextual Flexibility: Present Status and Future of Customary Marine Tenure in Solomon Islands. *Ocean & Coastal Management* 40: 253–69.

—— 2003. Both Sides of the Beach: Knowledges of Nature in Oceania. In H. Selin (ed.), *Nature Across Cultures: Non-Western Views of the Environment and Nature*. Dordrecht: Kluwer Academic, pp. 243–75.

—— 2015. The Western Solomons and the Sea: Maritime Cultural Heritage in Sociality, Province, and State. In E. Hviding and G. White (eds), *Pacific Alternatives: Cultural Politics in Contemporary Oceania*. Canon Pyon, Herts.: Sean Kingston Publishing, pp. 118–44.

—— 2016. Europe and the Pacific: Engaging Anthropology in EU Policy-making and Development Cooperation. In T. Bringa and S. Bendixsen (eds), *Engaged Anthropology: Views from Scandinavia*. New York: Palgrave Macmillan, pp. 147–66.

Igler, D. 2013. *The Great Ocean: Pacific Worlds from Captain Cook to the Gold Rush*. Oxford: Oxford University Press.

Jetnil-Kijiñer, K. 2017. *Iep Jāltok: Poems from a Marshallese Daughter*. Tucson, AZ: University of Arizona Press.

Lal, B.V. and K. Fortune (eds). 2000. *The Pacific Islands: An Encyclopedia.* Honolulu: University of Hawai'i Press.

Lazrus, H. 2012. Sea Change: Island Communities and Climate Change. *Annual Review of Anthropology* 41: 285–301.

Matsuda, M. 2012. *Pacific Worlds: A History of Seas, Peoples, and Cultures.* Cambridge: Cambridge University Press.

Mulrennan, M.E. and C.H. Scott. 2000. *Mare Nullius*: Indigenous Rights in Saltwater Environments. *Development and Change* 31: 681–708.

Nicholls, R.J. and A. Cazenave. 2010. Sea-level Rise and Its Impact on Coastal Zones. *Science* 328: 1517–20.

Nunn, P. 2009. *Vanished Islands and Hidden Continents of the Pacific.* Honolulu: University of Hawai'i Press.

Rudiak-Gould, P. 2013. *Climate Change and Tradition in a Small Island State: The Rising Tide.* New York and London: Routledge.

Tabe, T. 2016. *Ngaira Kain Tari – 'We are People of the Sea': A Study of the Gilbertsese Resettlement to Solomon Islands.* PhD dissertation, Department of Social Anthropology, University of Bergen.

Teaiwa, K.M. 2015. *Consuming Ocean Island: Stories of People and Phosphate from Banaba.* Bloomington, IN: Indiana University Press.

Thomas, N. 2010. *Islanders: The Pacific in the Age of Empire.* New Haven, CT: Yale University Press.

Tong, H.E.A. 2015. 'Charting its Own Course': A Paradigm Shift in Pacific Diplomacy. In G. Fry and S. Tarte (eds), *The New Pacific Diplomacy.* Canberra: Australian National University Press, pp. 21–4.

UN (United Nations). 1982. United Nations Convention on the Law of the Sea of 10 December 1982.

Vidas, D. 2011. The Anthropocene and the International Law of the Sea. *Philosophical Transactions of the Royal Society* 369: 909–25.

# 11. Islands of Hope and Despair
## Scaling the Collapses and the Collapse of Scales

*Frank Sejersen*

Throughout the world, indigenous peoples are increasingly struggling to navigate a world where issues of climate change add to the other problems of marginalisation, inequality and discrimination. Often, they are identified as particularly vulnerable, due to the fact that they are unable fully to participate in the political and economic systems of which they are a part. Indeed, indigenous people's position and the problems of inequality that affect them can be understood as the outcome of their exposure to globalisation in the form of climate change and the market economy (Leichenko and O'Brien 2008) on top of colonisation. In this chapter, I shall explore the question of inequality and justice by applying 'scale-making' as an analytical device. The concept of 'scale-making' is used to draw analytical attention to how people construct scales (for an elaboration, see Sejersen 2015); for example, how people scale what they conceptualise as local and global (see also Tsing 2012). In particular, I focus on how groups use scale-making to put forward moral claims. Based on two climate-related episodes where indigenous perspectives are present and made relevant, I analyse how anticipated collapses were creatively (re-)scaled and how other scales collapsed in that process. The episodes are demarcated on the basis of the intense activities that took place on two quite different Arctic islands: Kivalina in Alaska and Maniitsoq in Greenland. Although the islands are very different, the episodes mark the islands as sites where agency, responsibility and temporality are grappled with, imagined and evoked in order to understand the potential collapse of an indigenous community. The chapter argues that the islands can be approached as sites of intensive collective moments of reflection during the episodes where scales were re-visited by local inhabitants in order to mobilise hope and to pursue moral claims.

The two case studies have in common the fear and anticipation of imminent community collapse and the fact that action was taken to mobilise hope. The anticipated collapses resulted in intensive sense-making activities where community members tried to grapple with what took place. Community members worked creatively to rethink the scales in which the problems were addressed and interpreted. The episodes can be understood as moments of intensive collective scale-making. As a point of departure, scales are seen as particular space-time configurations produced and mobilised by communities of scale-makers. The conscious use of (re-)scaling is often linked directly to particular ideological and political positions, which make it possible to talk about the politics of scales. Thus, scales can be seen as an integral aspect of the structuring of discourse and of sense-making. In line with this, Erik Swyngedouw argues that 'the multiplicity of scalar levels and perspectives also suggests that scale is neither an ontological given and [an] a priori definable geographical territory nor a politically neutral discursive strategy in the construction of narratives' (1997: 140). Not only does the use of particular scales offer a frame for explanations and the production of contexts but it also carves out possibilities in which causal relations, events and agencies emerge. Subsequently, scale-making is a productive act when addressing responsibilities and when acts of claiming, shaming and blaming are pursued. Thus, scale-making is inherent in the construction and mobilisation of morality as well as political identities; scale-making indeed creates the foundation for the construction of power geometries. Swyngedouw points out that '[s]cale ... is both the result and the outcome of social struggle for power and control' (1997: 140). Although scale-making is inherent in all political work, the chapter is based on the idea that the negotiation or struggle over scales may be more intense on some occasions and that scale-making may carry transformative power.

But when do these moments of intensive collective scale reflection and negotiation emerge? According to Rebecca Bryant (2016), the perception of being in a 'time of crisis' carries a particular sense of present-ness (a special consciousness and awareness of the present) that is normally not part of one's everyday experience. Bryant argues that futures that are difficult to anticipate may produce this special sense of present-ness which she terms the *uncanny present*: 'The uncanny present ... refers to moments when the present that I normally do not perceive as such becomes anxiously visceral to us as moments caught between past and future' (2016: 20). This *in-between-ness* of the present endows it with transcendence and contingency, and the uncanny present emerges as a moment that can be understood as decisive for the future – a turning point. Moments of collective and critical reflections on how things, people, time and relations are joined together can be approached as critical thresholds and by doing so emphasise the sense that these moments, where the ordinary becomes something viscerally present, can be understood as both decisive and liminal, or outside ordinary time (Bryant 2016: 20). In such moments, we acquire, according to Bryant, 'a sense that what we do in this present will be decisive for both the past and the future, giving to the present the status as a threshold' (2016: 20). In her development of the idea of an uncanny present, Bryant gets inspiration from Freud, who used the term 'uncanny' to refer to the ordinary and familiar made strange. A situation in which the familiar may provoke terror requires attentiveness, alertness and action. Because of this familiarity, we have difficulties separating ourselves from it. In such moments of crisis, the present emerges as uncanny because of the inability to anticipate the future and because temporality itself becomes weightier as it bears the burden of a future that seems difficult. Immediacy is infused in people's imagination, fostering an awareness of the present (in its present-ness) as a decisive node between the past and the future. In these difficult moments, the present determines the future in people's minds. The normal links that people make to tie past, present and future in order to allow a sense of anticipation disappear. Imagination (as a mode of action) can – in these situations of uneasy temporal relations – be mobilised and centred on the idea of threshold. Bryant (2016: 23) argues that such a threshold implies the crossing into another space of time and a radical reorientation of the present. I take this idea of crossing to apply it to acts of re-scaling, because such activities involve the reflection on and creation of new space-time configurations. The collective aspect relates to the social practising of *re*orientation, *re*configuration and *re*flections on space-time configurations. At the two islands, this involves intensive collective processes and the islands become sites of scale reflection and scale-making in different ways, and the communities mobilised are also quite different in scale. However, they share the understanding of the present as uncanny and they try to grapple with the uncertainty of the future and potential ways out of a perceived collapse.

## KIVALINA

Kivalina is a small, remote native village located on an island (an 8-mile-long ocean spit) made of sand and gravel in northwest Alaska. The majority of the 450 inhabitants are Iñupiat, well known for their bowhead whale hunt, which they pursue like many of the other Iñupiat villages along the northern coast. As part of a programme to control the native peoples, the Bureau of Indian Affairs constructed a school at the tip of the island in 1905. It can be seen as an active act of stimulating a relocation of the native population in the area. The native people started to migrate to the location, which some of them had used earlier for temporary hunting camps. When the post office was established in 1940 and the airstrip in 1960, its status as a permanent village was further consolidated. In the 1950s, the village became known to the American public because it was one of several native communities to be affected by experimentation with the peaceful use of atomic explosives. The plan was to detonate an atomic device, some 100 times bigger than the Hiroshima bomb, close to the village in order to establish a harbour from which the region's wealth in oil and minerals could be shipped out. Project Chariot – as it was called – was also intended to benefit the local population which was to be relocated to the new harbour site (Chance 1990: 143–7; O'Neill 1994). In 1962, after several protests from the native peoples, the US Atomic Energy Commission announced that Project Chariot would be 'held in abeyance'. However, relocation is a recurrent theme for the village.

Today, the *Kivalinarmiut* (persons from Kivalina) are themselves asking to be relocated. Seen from their position on the shores facing the Chukchi Ocean, the village is facing collapse. Storms are eroding the coastline and with each bite the waves take, the island is getting smaller and the village is undermined. The village has been asking for a relocation programme since 1992 (Shearer 2011: 114), but a severe storm in 2004 made the need and urgency apparent to all members of the community.

> What had once been a normal occurrence of annual fall sea storms became a life-threatening event. As the stormy days progressed, the people became concerned over the amount of land falling into the ocean. The island seemed to be falling apart and disappearing into the Chukchi Sea before the very eyes of its inhabitants. Volunteers from the village began to work feverishly to hold the island together but every effort, every object placed along the edges, was being sucked into the angry sea. The volunteers worked through the pitch darkness of the cold fall night ... trying to save the people of Kivalina and the island. Evacuation by air was not an option because of the weather conditions and because the village was surrounded by the rough waters of the sea storm. This meant evacuation was also not an option by boat to the mainland. There was nowhere to go and nothing the volunteers did worked to keep the island together – the people were trapped. (Kivalina Tribal Administrator Colleen Swan cited in Shearer 2011: 100)

*Kivalinarmiut* argue that the erosion problems have worsened due to the increased number of storms, and because the sea ice no longer forms and acts as a protective shield against the destructive sea. These changes have left the village exposed and vulnerable. Eventually, the village will collapse and the community is struggling to understand the predicament and to think of ways to solve the problem. Kivalina Tribal Administrator Colleen Swan (cited in Shearer 2011: 12) expressed the community's frustration and concern in the following way: 'Where will we go, and who will help us move?'

Even though one could say that the cause of the problem is easy to identify (waves threatening a vulnerable community), *Kivalinarmiut* are trying to make sense of the situation and to create a framework within which different causalities can emerge. 'Climate change' is the overall explanation of the predicament, which is given by community members as well as outsiders. Once again, Kivalina and its need for relocation has become the focus of media attention, as it was when it was planned to relocate the village as a consequence of Project Chariot. Today, the collapse of the village is seen to be linked to global climate change and this link has made the village's plight central to our understanding of the effects of climate change. Inuit are often seen as iconic representatives of climate change, witnesses on the ground, proving the abstract warnings of scientists (Martello 2008). Kivalina has become instrumental in scale-making activities in the climate change discourse and used to produce ideas of 'tipping points' where 'the global' and 'the local' are seen as related.

If scaled as a problem emerging as a result of the sea threatening a vulnerable community, the solution becomes one of protecting the village from the encroaching sea. The US Army Corps of Engineers has designed and established protective constructions, but they are only considered temporary. However, when the village collapse is scaled in a way that links the erosion of the island to global greenhouse gas emissions, industries emerge as a crucial driver of the concrete problems experienced on the shorelines of Alaska and thus become part of the solution as well.

## Creating a Scale of Hope

In an attempt to raise funding for a resettlement project, the village filed a lawsuit against 24 oil, gas and electric companies in 2008 (Native village of Kivalina *vs.* ExxonMobil Corporation, et al.; see US District Court for the Northern District of California Oakland Division 2009). In doing so, the villagers tried to make the companies fiscally responsible for the emissions of greenhouse gases that are contributing to the warming of the Arctic. In fact, blaming specific subjects can be seen as a new way of adapting to climate change. This mobilisation of such an interpretative framework to understand the village collapse is an act of creative scale-making which created new hope in the community, at the time. The emergent community collapse was presented as accelerating: urgent action was required. According to Kivalina's attorney: '[t]he village is being wiped out by global warming and needs to move urgently before it is destroyed and the residents become global warming refugees ... It's battered by winter storms and if residents don't get some money to move, the village will cease to exist' (cited in D'Oro 2010).

This scale-making activity drew capitalist enterprises and industrial expansion directly into the picture. The particular scaling evoked through the lawsuit made it possible to establish a framework in which certain ideas of responsibility and accountability could emerge and be made to operate and thus also point at what the scale-makers considered to be the causes of injustice and moral failure. The lawsuit was based on the common law theory of nuisance: Kivalina claimed monetary damages from the industry because it saw the industry as the cause of the community collapse. However, the court dismissed the case as the issue of regulating greenhouse gases was considered a matter for policy-makers rather than the court system and because Kivalina's injuries were not 'fairly traceable' to greenhouse gases emitted by the particular defendants. Kivalina's scaling project, which established the framework for their moral claim, was not in concordance with the court's framework. Even though it is possible to see the issue and the connections within this new scale, the existing institutional 'tools' make it impossible to uphold the case. According to Judge Armstrong, the courts must ask whether they have the legal means to reach a ruling that is principled, rational and based upon reasoned distinctions (US District Court for the Northern District of California Oakland Division 2009: 10). Particular principles, rationalities and distinctions can thus be seen as made possible or impossible by certain scaling practices. When Kivalina challenged the courts by establishing a scale where global warming, village collapse and industrial companies can be encompassed within the same framework, it also challenged the scales upon which the court's principles, rationalities and distinctions depend and from which the court's moral rulings arise.

The claim was appealed several times but was effectively closed by the Supreme Court in 2013. With inspiration from Swyngedouw (1997: 141) the court case can be seen as an illustration of 'the struggles between individuals and social groups through whose actions scales and their nested articulations become produced as temporary standoffs in a perpetual transformative sociospatial power struggle'. Kivalina's legal claim, and the way the village inscribed causality and legitimacy onto it, worked on a scale that could not be dealt with by the court, and Judge Armstrong could therefore conclude that the 'Clerk shall close the file and terminate all pending matters' (US District Court for the Northern District of California Oakland Division 2009: 24). Kivalina's court action is not necessarily a case in which a weak party is overrun by a strong agent. I would rather say that the case shows that Kivalina's legal action was based on a particular understanding of causality that was made possible by the evocation of a particular scale, which had no resonance in the existing juridical (and thus moral) regime. Its scale-making activities, which infused meaning and made it structurally possible to relate the village erosion to industrial pollution, collapsed when it was tried in a system where causality and responsibility work within a different framework.

Kivalina's lawsuit directs our attention to the destructive aspects of modern consumption patterns and lack of proper regulation to mitigate negative effects. When this is made centre stage, the industries, consumers and indigenous peoples are tightly bound together in an interconnected system. In this scale configuration new understandings of justice and injustice, as well as responsibilities, emerge clearly. However, the legal system found it difficult to blame a particular legal agent. Climate change, like the world economy, has a transcendental aspect because it works on several space-time scales simultaneously. In general, court cases linked to relations between production, consumption and pollution are extremely slippery, even though causalities are recognised.

The Kivalina case shows how a community mobilises hope by arguing within a new scale in order to infuse a new moral order in a present where a collapse is emerging. The case also shows how this reconfiguration meets resistance and is hard to accept by the existing systems working with other hegemonic space-time configurations.

## MANIITSOQ

The mountainous island of Maniitsoq on the west coast of Greenland is the home of approximately 2,500 inhabitants, primarily Greenlanders of Inuit descent. The Danish colonial trade established a community there in 1782, in order to support and expand Danish whaling activities. Later, the town of Maniitsoq – which is the only settlement on this island of 150 km$^2$ – grew into an important centre for industrial fishing from about 1920 due to the presence of large amounts of cod attracted to the region by the warming seas.

## The Emerging Status as Centre

The in-migrating Greenlanders, who wanted to partake in the fish bonanza, gradually underpinned the town's status as a fishing centre. Increasingly, fishing replaced seal hunting as the most important economic activity. As a result, Greenlanders started to invest in small wooden boats and to give up the large skin boats. The production chain of the emerging fishing industry was able to offer job opportunities to persons with a variety of skills and became a very inclusive industry in which women and men of all ages could participate. Both in the 1930s and later in the 1950s to 1970s, Maniitsoq experienced a fishing boom and the town acquired a significant role in developing the modern Greenlandic cod fisheries. Thus, the town played an important role in the Greenlandic nation-building process. The history of Maniitsoq town and the national history of the fishing industry go hand in hand. The use of the national workforce and national resources co-produced the nation-building process in Greenland. However, in the 1980s the fishing sector overheated because of overcapacity. Since then, the sector has been undergoing a process of structural rationalisation and optimisation, which has resulted in the reduction of jobs and vessels. Furthermore, the sector has experienced new ways of regulating the fisheries that have resulted in the concentration of benefits for fewer people. The productive fishing industry is paradoxically contributing a significant part of the country's income, on the one hand, and offering fewer and fewer job opportunities, on the other.

## The Anticipated Collapse

Over the years, the town's population has fallen dramatically. People are looking for jobs elsewhere in Greenland. Increasingly, people and companies move away from Maniitsoq, and many *Maniitsormiut* (people from Maniitsoq) are concerned that this brain drain has brought the town to a critical tipping point where it becomes difficult to maintain the dynamics of the community. The contemporary national 'race for limited resources' in Greenland has also created a competitive atmosphere between Greenlandic towns fighting to attract and hold on to human and non-human resources. *Maniitsormiut* anticipate a community collapse if the present drain on resources is not reversed. Locally, people are struggling to understand the dynamics and reasons for this anticipated community collapse.

Often, people in Maniitsoq link the feared collapse directly to policies of redistribution of national resources and attention, which seem to favour centralisation in only a few towns at the expense of many other communities (Sejersen 2007). Locally, the driver of the town's collapse is not linked directly to local understandings of decreasing resources in Maniitsoq ('there is plenty of fish') or to the economic system logics and consequences of neoliberal rationalisation ('fishing should be profitable'), but to resource allocations, prioritisation and planning by the national authorities. Political mismanagement and bad judgement and accelerating national inequality are recurrent themes (Sejersen 2015). People in Maniitsoq vigorously oppose centralisation, and by doing so they hope to get their 'share' of the available national resources in Greenland. The dominant way of thinking in Maniitsoq is to link the anticipated community collapse to national dynamics. Often, arguments to reverse the situation are formulated as a moral claim to be (re)integrated into the national domain, and thereby to be (re)installed in the national flow of resources.

Thus, the understanding of the anticipated collapse points to causalities, agents and employments that reflect and are made possible by a certain way of scaling the problem. It is by pursuing a productive creative act of engaging local dynamics with national dynamics that local scale-makers make it possible to translate the local predicament into a national political and moral question whereby some specific agents can be blamed for creating national inequality and uneven national development.

## The Emerging Solution

Quite unexpectedly, Greenlandic authorities were approached by the world's second biggest aluminium producer (Alcoa) in 2006 and asked whether it was possible to construct an aluminium plant, to be run on cheap energy produced by hydroelectric facilities using the water from the melting glaciers near Maniitsoq. The warming Arctic has accelerated the disappearance of sea ice and glaciers, and Greenland has attracted growing interest from foreign companies, which are looking for new business opportunities. Narratives of climate change have produced a discourse of the Arctic as a new economic frontier and the region has emerged as a promising new scene for investments (Sejersen 2015). These new opportunities were soon used by the *Maniitsormiut* as a way of creating new resources to 'be harvested', making it possible for them to escape the national zero-sum politics of Greenlandic resource redistribution. Maniitsoq's collapse was suddenly perceived to be avoidable as a result of the infusion of outside agency and money. Greenland in general and Maniitsoq in particular started to reorient their understandings of their position and possibilities in an economy that was not fixed to existing national sectors and the limited national marine resource base but to large-scale industries navigating the global financial markets. This reorientation and re-scaling prompted new dynamics of place-making and space competition at the local and national level (Sejersen 2015). Now, places were reinvented and reinterpreted to match the expectations of outsiders. New place configurations emerged as different trajectories were created (Massey 1994).

*The Emerging World*

Alcoa wanted to construct a massive industrial complex consisting of a 1.5-km-long smelter and two large-scale hydroelectric facilities. It soon became apparent that there was not sufficient skilled labour in Greenland, and that Greenland had to import workers in order to match the scale of the project. Thus, several thousand foreign workers were expected to construct the facilities over a period of approximately five years – and then leave the country. The plan was that Greenlanders were to have the jobs in the aluminium smelter once production started. For Greenland, this smelter project was believed to be a game-changer for the national economy and a tool to close the widening and devastating abyss between public income and public expenditure. Greenland had long been warned about its problematic economic situation, and it was repeatedly made clear to the public that the 'status quo is not an option' (for example, by Hammond cited in Pacini 2014) and by the Taxes and Welfare Commission (Skatte og velfærdskommissionen 2010: 6). Wide-ranging structural reforms were believed to be the only way out of the economic predicament. The Alcoa project gave new hope and triggered an energetic rethinking of society (Sejersen 2015). The proposed smelter can be seen as a 'hope technology', to use Sarah Franklin's concept (1997).

Greenland had a very positive attitude to the idea of admitting thousands of (presumably) Chinese workers. This foreign labour force was expected to work under the rules, laws and agreements applicable to the Greenlandic labour market (Sejersen 2015: 175). Greenland had firmly embedded the morality of the labour market on a national scale. The history of the labour market in Greenland has its own colonial legacy, which framed the moral discussion in a particular way. For Greenlanders, equal rights and salary for the same work is a milestone in the decolonisation process; ethnic background should not determine the level of salary. Until the 1990s, persons with a Danish background were able to receive a higher salary, and the elimination of this system is considered by many Greenlanders to have been a great step forward in severing the colonial relations of inequality in which they formerly found themselves, offering a new way to practise national self-determination. In 2007, the Greenlandic government therefore made it clear to Alcoa that foreign workers were to work under the same rules and agreements as the rest of the workforce in Greenland. This idea reflects an act of scaling which reinforces national control and benefit. It gives the project a particular national configuration and it provides agency to the national authorities. The moral contracts invested in the labour market as national were to be maintained when the country engaged in new large-scale activities in cooperation with foreign companies wanting to operate in Greenland.

However, in 2011, Alcoa challenged the national narrative and framework, which was used by the Greenlandic authorities to understand, promote and control the project. The company made it clear to the Greenlandic authorities that the projected construction had to be 'internationally competitive'. Suddenly (seen from a Greenlandic point of view), the logic of the project and its legitimacy were entangled in an up-scaled framework where the ongoing negotiations of value, morality and benefits were to take place. Alcoa pointed out that the need to make the construction internationally competitive was to make it worthwhile when compared to other aluminium projects elsewhere in the world. Practically speaking, the act of up-scaling the project's framework provided a new comparative framework in order to infuse new logics. Alcoa used the up-scaled project frame to emphasise that overheads had to be reduced substantially in order for the project to be competitive. This means that the idea of a 'globally competitive marketplace' is interposed into the discussions and evoked as 'part of the game', which is to drive decisions and priorities. Thus, Greenlanders found themselves entangled in a situation where the company skilfully evoked the 'global' as the hegemonic scale by means of the idea of competition in a trans-boundary international marketplace, where both *labour* and *place* became competitively interlinked parameters to evaluate cost and value. In this new framework, labour and place were put together in new ways. Alcoa proposed to make the project internationally competitive by reducing construction costs. That strategy could be made possible if Alcoa were allowed to use non-Greenlandic labour, who were to be paid what the company termed 'international wages' which are below the minimum wage in Greenland. Thus, if Greenland wanted to emerge as an attractive, internationally competitive location, the authorities had to change national rules and regulations to admit a low-paid international workforce.

Furthermore, Alcoa argued that the use of technology and labour should be governed by regulations effective in *other* countries with *similar* large-scale industrial projects (Naalakkersuisut 2010: 94). Allegedly, such a new framework would reduce the costs by US$2.5 million (Benson and Sørensen 2011) but would also require – according to the Greenlandic government – that the representatives of both the labour market and the Parliament contribute to the 'creation of the required legislative framework' (Naalakkersuisut 2010: 7). When Alcoa successfully established a framework of a 'global competitive marketplace', Greenland not only had to manoeuvre within a scaled space of international capital flows and world market possibilities, but also *actively* to relate to framework conditions present elsewhere in the world. Greenland had to rethink its market value and to understand new (global) dynamics, which increasingly define the value of Greenlandic resources. The influx of foreign investments would only be possible if national rules and regulations on the labour market were changed in such a way that the circulation of capital and labour would be freed from national moral imperatives, with the purpose of ensuring the accumulation and circulation of capital in a fully integrated global company economy. Thus, Greenland had to establish a new situation and self-perception of societal coherence that was fit for the new global actors. Greenland had to emerge as a place for investments that was as good as elsewhere. According to Neil Smith, as reviewed by Marston (2000: 229), '[t]his equalization is accomplished through the universalization of the wage-labor relation through both formal and real spatial integration into a global system'. Imported workers are, with reference to the mechanisms of 'global competition' and 'international wages', expected to work under 'special' regulations, and for 'special' wages that differentiate them from the national workforce.

In many countries, it is accepted that a particular kind of imported labour force works in a 'state of exception' from the regulations in place for other workers in the countries in question. To put this more precisely: the framework for labour in a particular country is changed in order to allow this kind of labour mobility and thus make the 'state of exception' a more or less permanent possibility. In Greenland, many criticised such a step. The president of the Greenland Labour Union (SIK), Jess G. Berthelsen, quickly issued the following statement:

> I am warning the [Greenland] government against opening up for low-paid Chinese. Foreign workers are to have proper salaries. No firm should try to bypass the existing agreements in Greenland. This is a pre-set ticking bomb. If we say 'yes' to Alcoa, then other companies

will line up and demand the same meagre agreements when oil production and mining are set in motion. (Benson and Sørensen 2011, translated by the author)

The president of the Employers' Association of Greenland (GA) was also sceptical, primarily because such 'special' arrangements would benefit only very few companies and thus create a competitive distortion in Greenland (Leth 2011). Both spokespersons suggest that national problems could emerge if outside companies were assigned a privileged position due to the new up-scaled framework.

## Collapse of the National Labour Market

When the place (Greenland), the project (aluminium smelter) and the labour (skilled workers) are scaled within the framework of a 'global competitive marketplace', it gives rise to certain subject positions, responsibilities, potentials and limitations for stakeholders, governments and companies. The Greenlandic Parliament had to create a new legal framework, to rethink the country's future labour market situation and to differentiate between businesses. That requirement emerged out of their aspirations for a future national economy based on the large-scale industrialisation of the country fuelled by foreign investments and engagement. New resources to boost the Greenland economy thus required a political and moral manoeuvre of up-scaling. In 2012, after long debates in the Greenlandic Parliament focusing on problems with social dumping among other issues, a new set of regulations for large-scale industrial projects was passed to create the necessary legal framework – the so-called *Large Scale Projects Act* (Storskalaloven – 'Lov for Grønland om udlændinges adgang til opholds-og arbejdstilladelse i anlægsfasen af et storskalaprojekter').

Basically, the Large Scale Projects Act allows companies to use cheap foreign labour. This process can be understood as an external pressure on Greenland to allow its labour market to operate in a manner that is unfettered by national regulations and agreements in order to create a gateway for the mobility and circulation of a transnational low-paid labour force. Such a view is expressed by Leichenko and O'Brien (2008), amongst others, when they argue that people are *exposed* to economic globalisation. It is within this global economic structure of causalities and feedback loops that winners and losers are produced. However, this opening up and reformulation of the national place can also be seen as part of a process whereby Greenland itself is actively co-producing, evoking, enacting and mobilising the scale and the logics of global integration and mobility. The 'global' and 'global processes' cannot simply be located as 'outside' Greenland: they are a force penetrating the state and the local (for a discussion, see Sejersen 2015). It is rather the case that 'global' and 'global processes' are enacted along a trope of the 'global as a competitive marketplace'. Greenland actively scales, imagines, relates to and alters the framework in which it has to navigate and maintain what are considered to be the 'proper' roles and relations of governments, unions, workers and companies. The 'global' and 'globalisation' are installed as important aspects of Greenland's own vision for an economically viable future. It is within this new geometry of power and global place-making that the authorities have to reconfigure national self-perception and morality.

Greenland's desire for international economic attention and the construction of large-scale projects which could underpin more economic dependence from Denmark meant that the country's labour market, understood and scaled as national, collapsed. The engagement with a global firm challenged and led to a reformulation of the moral and political understanding of the legitimacy of the social contract between state and citizen within a national framework and the national labour market. Greenland has always been dependent on world market prices, global consumer requirements and economic dynamics in spheres, places and scales that lie outside the control of Greenland. However, the *Large Scale Projects Act* adopted by the Greenland Parliament is different, as it introduces new national legislation, which reformulates national self-understanding and the logics of the national labour market in order to mobilise a workforce at another scale, introducing new dynamics into national economic development and increasing Greenland's competitive advantage. The *Large Scale Projects Act* can be seen as a new space-time configuration, which infuses hope and potentiality into a national and community economy facing collapse, but it also redefines and challenges the national at the very same time. The case shows how the existing scale of the labour market has been fundamentally changed in order to avoid anticipated collapse and reveals the extent to which this transformation is institutionally backed. For Greenland as a nation, and Maniitsoq as a community, the *Large Scale Projects Act* can be seen as an intensive and decisive moment of scale reflection that mobilised new ways of seeing one's position in the world economy in an era of climate change where the Arctic had become more attractive to commercial interests. However, it was also a productive activity of scale-making that positioned Greenland in a new moral order.

## CONCLUSION

Scale-making is a political act and is also used as such by indigenous peoples. Re-scaling can be a powerful way to break isolation and inscribe people and places with new meanings in new moral relations. Such an act may initiate hope and fuel future imaginaries by opening up horizons where times yet to come are made possible. Thus, scale-making is not only related to degree and extent but also to kind and content. The people of Kivalina, facing the destruction of their town, pursued a productive act of up-scaling to attract attention and financial aid by turning a small town into a community arguing for justice and compensation on the basis of causal relations on a global scale. However, the produced scale collapsed as the village's actions met opposition by the established system of morality upheld by the courts. It was impossible for the court system to handle the proposed moral claims made possible by the town's up-scaling activities. Maniitsoq and Greenland, facing financial problems, engaged in processes of negotiation with a globally oriented company in order to attract outside resources and to avoid community collapse. However, the flow of 'outside' resources into the national sphere was only possible if certain national regulations (from the outside seen as idiosyncratic national barriers) were removed in order to make the Greenlandic labour market more aligned with the requirements of a more 'international' understanding of 'international' economic relations. It resulted in a reconfiguration of the labour market that was formerly understood to be guided by a national moral imperative. The Greenlandic authorities (re-)scaled its understanding of the labour market through the passing of new legislation to make the influx of low-paid foreign workers possible in an attempt to avoid economic collapse. In doing so, they changed the existing moral codex of the labour market which had been perceived as a hallmark of post-colonial indigenous emancipation and part of the nation-building process in Greenland. Thus, the idea of a national labour market underpinned by a post-colonial moral imperative solely defined by Greenland collapsed.

Both cases can be seen as intense moments of collective scale reflection carrying moral momentum and as thresholds demarcating a new scene of potentiality. It is important to notice that the potentiality in up-scaling is created by the people facing problems. Up-scaling and the reframing of moral relations were perceived as creating new resources. The two cases are not thought of as examples of how small-scale community perspectives are overrun by large-scale global hegemonic interests. Rather, the cases show that re-scaling (in these cases, up-scaling) can be seen as productive and can underpin people's societal vision. The cases also show that something is lost as well as gained in these manoeuvres of re-scaling.

### REFERENCES

Benson, P. and B.H. Sørensen. 2011. Alugigant Kræver Billige Kinesere i Grønland. Business.dk, 11 March. Available at: www.business.dk/industri/alugigant-kraever-billige-kinesere-i-groenland (accessed October 2018).

Bryant, R. 2016. On Critical Times: Return, Repetition, and the Uncanny Present. *History and Anthropology* 27(1):19–31.

Chance, N. 1990. *The Iñupiat and Arctic Alaska. An Ethnography of Development*. Fort Worth, TX: Holt, Rinehart and Winston.

D'Oro, R. 2010. Eroding Alaska Village Appeals Lawsuit's Dismissal. *The Seattle Times Online*, 28 January. Available at: https://tinyurl.com/ychhg64j (accessed October 2018).

Franklin, S. 1997. *Embodied Progress: A Cultural Account of Reproduction*. London: Routledge.

Leichenko, R.M. and K. O'Brien. 2008. *Environmental Change and Globalization: Double Exposure*. Oxford: Oxford University Press.

Leth, H. 2011. Det er Ikke Særordninger men Handlekraft, der er Brug for. Greenland Business Association, 8 March. Available at: https://tinyurl.com/y8g4ugmr (accessed October 2018).

Marston, S. 2000. The Social Construction of Scale. *Progress in Human Geography* 24(2): 219–42.

Martello, M. 2008. Arctic Indigenous Peoples as Representations and Representatives of Climate Change. *Social Studies of Science* 38(3): 351–76.

Massey, D. 1994. *Place, Space and Gender*. Minneapolis, MN: University of Minnesota Press.

Naalakkersuisut. 2010. *White Paper on the Aluminium Project Based on Recent Completed Studies, Including the Strategic Environmental Assessment (SEA)*. Nuuk: Greenland Self-Government.

O'Neill, D. 1994. *The Firecracker Boys*. New York: St. Martin's Press.

Pacini, N. 2014. Greenland Hopes for Future in Mining, Independence from Denmark. *Medill News Service*, 29 September. Available at: http://dc.medill.northwestern.edu/blog/2014/09/29/pacinigreenland0924/ (accessed October 2018).

Sejersen, F. 2007. Indigenous Urbanism Revisited – The Case of Greenland. *Indigenous Affairs* 3: 26–31.

—— 2015. *Rethinking Greenland and the Arctic in the Era of Climate Change. New Northern Horizons*. London: Routledge.

Shearer, C. 2011. *Kivalina, a Climate Change Story*. Chicago, IL: Haymarket Books.

Skatte og velfærdskommissionen. 2010. *Hvordan Sikres Vækst og Velfærd i Grønland?* Nuuk: Grønlands Selvstyre.

Swyngedouw, E. 1997. Neither Global Nor Local: 'Globalization' and the Politics of Scale. In K. Cox (ed.), *Spaces of Globalization*. New York: Guilford, pp. 137–66.

Tsing, A. 2012. On Nonscalability: The Living World is Not Amenable to Precision-nested Scales. *Common Knowledge* 18(3): 505–24.

US District Court for the Northern District of California Oakland Division. 2009. Native Village of Kivalina and City of Kivalina, Plaintiffs, *vs.* Exxonmobil Corporation, et al. Defendants. Case no. C 08 – 1138 SBA. Available at: www.courtlistener.com/opinion/1414248/native-village-of-kivalina-v-exxonmobil-corp/ (accessed October 2018).

# 12. Using a Glacier Website to Promote Action and Build Community
## Engaged Anthropology in the Digital Age

*Ben Orlove, Kerry Milch and Laura Uguccioni*

The interconnected environmental and economic crises in the contemporary world create opportunities for what many have termed engaged anthropology. The website GlacierHub explores one such opportunity. The activities of the website since its establishment in 2014 show the operation of engaged anthropology during this time of crises. It should be noted, at the outset of the discussion, that both topics – the crises and engaged anthropology – are recognised to be broad, and that they elude precise definition. One major review of engaged anthropology (Low and Merry 2010) emphasises engaged anthropology as 'practice' (2010: 203), contrasting it with research that is exclusively academic in nature. It noted that engaged anthropology 'addresses public issues' (2010: 203) offering a second contrast with activities that are confined to the internal workings of a discipline. Linked to these two points, the review notes that engaged anthropology 'respects the dignity and rights of all humans and has a beneficent effect on the promotion of social justice' (2010: 204).

In this chapter, we discuss how this website provides an example of engaged anthropology in the specific domains of climate change and glacier retreat. We argue that websites and social media like GlacierHub can increase awareness of environmental crises and build concern about them. Moreover, they can support communities that confront these crises, and promote action to address them.

### AN OVERVIEW OF ENGAGED ANTHROPOLOGY

Though antecedents of anthropology can be found in the remote past and across many different social settings, the modern discipline of anthropology began to take shape in the nineteenth century. It emerged from a number of sources. Some of its intellectual origins lie in the encyclopedic and classificatory impulses that shaped a great deal of writing in that century, but it has also been deeply intertwined since its inception with the environmental and economic systems that are now in crisis. The societies and cultures that were the objects of early anthropological inquiry were available to researchers because of the expansion of empires, linked directly or indirectly with capitalism. The first generation of ethnographic researchers had a variety of engagements with these empires, at times directly supporting their growth and facilitating the domination of the societies which they encountered, at other times challenging the terms, if not the basic fact, of this domination. Even in early nineteenth-century Britain, when the nation's colonial empire in Africa was expanding, anthropologists were located on both sides of the abolitionist movement that sought the end of chattel slavery (Kuklick 2009). Anthropological engagements with public life and with politics in the twentieth century have had some high points, familiar to many anthropologists in the United States. Margaret Mead was a major public intellectual. She built on her field research in the Pacific in the 1920s and 1930s and her understandings of other cultures to reach large public audiences in the post-war decades. Her books and articles in major weekly magazines addressed issues related to gender and kinship (women's rights, child-rearing, birth control) and other issues of the time (nuclear war, world hunger, environmental degradation). Another major moment occurred in the 1960s, when anthropologists played a central role in the opposition to the US war in Vietnam by organising teach-ins, events modelled on the labour movement's sit-in strikes, in which they explained the war, and their opposition to it, by drawing on anthropological insights into the dynamics of state societies. These teach-ins became a central feature of anti-war social movements.

Others (Lamphere 2004; Checker 2009; Mullins 2011) have traced the history in greater detail, bringing it forward to the present. Some features of engaged anthropology have grown more complex in the present century. The end of the Cold War, the global awareness of terrorism, and the migrant and refugee crises have complicated the question of the public which engaged anthropology addresses, since national publics are becoming problematised, if not eroded, while global publics are emerging only incompletely (Goldstein 2014). The social movements which engaged anthropology show signs of fragmentation into smaller movements; though marked by alliances and intersectionality, they remain more difficult to locate. Where earlier engaged anthropologists could identify with broad social causes, contemporary ones often find narrower forms of advocacy and activism. The mass audiences that Mead and others addressed have also become divided, a shift facilitated by the rise of the Internet and social media and the decline of mainstream print journalism.

In this context, engaged anthropology itself displays multiplicity and fragmentation. Low and Merry (2010) note seven avenues for engaged anthropology: participating in policy-making, addressing social problems, connecting with media, serving as witnesses of violence and social change, co-producing knowledge with communities, supporting social assessment and ethics, and creating solutions (2010: 203). These overlap with six forms of engagement: 'sharing and support, teaching and public education, social critique, collaboration, advocacy, and activism' (2010: 204). Of these, social critique has occupied a great deal of attention, as anthropologists have debated whether critique itself constitutes a form of action (Osterweil 2013). Low and Merry note dilemmas as well: efforts at collaboration between anthropologists and others can lead to confrontations with existing power structures. On some occasions, inequalities persist because anthropologists benefit from their privileges as citizens of wealthy countries or as educated professionals. In other cases, anthropologists risk cooptation by dominant institutions. And on other occasions, it can be difficult to apply universal values – human rights, justice – in complex, fragmented political contexts, where the 'promotion of social justice' does not provide a clear road map of which forms of action to undertake, and which to avoid.

## A SPECIFIC CASE OF ENGAGED ANTHROPOLOGY: A GLACIER WEBSITE

The website GlacierHub, which launched on 7 July 2014, focuses on glaciers. The impulse to develop the glacier website came from a sense of urgency about climate change. The worldwide retreat of glaciers provided an opening to work on addressing climate change. Some earlier research on the history of climate change movements suggested that the greater presence of local and indigenous voices from the Arctic and from small island states had brought those areas forward in global climate change agendas, strengthening action on climate change overall, while a weaker presence of such voices from mountains and deserts limited the attention to those areas (Orlove et al. 2014). Therefore, GlacierHub aims to support the voices of local and indigenous people affected by the loss of glaciers.

The statement that appeared on the home page to introduce the website states:

We seek to expand and deepen the understanding of glaciers around the world. We tell stories of people who live near glaciers or visit them. We provide information about current scientific research about glaciers and we offer accounts of communities and organizations who are working to address the challenges brought by glacier retreat, whether through activism, policy and economics, through art, photography, or other means. (GlacierHub 2018)

A longer statement of purpose appears on another page, entitled 'About GlacierHub'. It emphasises the connection of glaciers to water resources and to natural hazards, and, in a determined effort to move beyond elements closely tied to economic valuation, it added 'glaciers are precious as well for their transcendent beauty'. To indicate the additional goal of building community, through social networks and social movements, it states that the website 'serves as well as a nexus to link people who are concerned about glaciers, so that they can communicate with each other and develop responses to the changes in glaciers'. The inclusion of concrete activity is signalled in the final sentences. This longer statement ends with a paragraph,

And glaciers are endangered. In all areas of our warming world, they are shrinking, as winter snows are no longer sufficient to replenish their melting. So glaciers can become a theme for people who are trying to make sense of our changing world. As people search for ways to comprehend and address climate change, glaciers often come forward as potent elements in thought and action. (GlacierHub 2018)

The content of the website reflects this statement, which has remained unchanged since the website's first appearance. The category with the largest number of posts is scientific research. Other posts fall broadly under the categories of policy and politics, community perceptions of glacier retreat, adaptation efforts, climate policy conferences, climate demonstrations, activism around issues of mining and the like. The website also has many posts on art, understood broadly to include not only established forms, particularly painting, sculpture and photography, and newer types such as installations, but also more popular genres such as film, television and video games.

These topics are linked by their connection with glaciers and by the website's stated purpose of disseminating information and building community. One important aspect of the website is the international nature of the authors. The majority of the pieces have been written by students at Columbia University, a number of whom come from countries outside the United States, including India, China, Nepal, Singapore, Peru, France and Australia. The website has featured posts written by researchers, community members, activists and journalists from Peru, Bolivia, Nepal, Kyrgyzstan and Iceland, as well as interviews with people from more than a dozen countries.

The website, developed in conjunction with anthropologists and psychologists at Columbia University's Center for Research on Environmental Decisions (Gertner 2009), rested on an understanding of the links which connect information, action and community. In this view, dissemination can create awareness, a term which conveys that people have not only received information, but that it plays a role in shaping the ways in which they see the world; they are prepared to recognise the importance of mountain glaciers, of people who live in mountains and of climate change, both in mountain regions and elsewhere. Concern and preparation move from the realm of cognition to realms of affect and volition. The awareness of glacier issues leads to worry about them, an involvement with them and a wish to do something about them, followed by concrete resolve. Action – the final word in the longer statement of purpose – moves beyond this intention to actual conduct or undertakings to achieve some aim, whether as an individual or in coordination with others. The term 'action' is broad, since it can take many forms; it is elusive too, since there is also a variety of ways that actions can be connected to glaciers and to climate change. The acts of individuals and groups would lead to community, another difficult term. Its suggestion of people linked by shared identities and cultural forms is often slippery and vague, and its implications of solidarity and mutual support, appealing as they are, can often mask elements of inequality and domination. As with the term action, this term indicated a hope, as well, that the website would serve to link people who would discover commonalities, share thoughts and undertake joint activities: hence the word 'hub' in the title.

These terms – awareness, concern, action, community – overlap with the elements of engaged anthropology discussed earlier. They correspond broadly to the long-established notion of social movement, which is linked to ideas of the public sphere as a space of communication and of action, connected broadly to the rise of capitalism (Habermas 1989). Though they are not tied directly to the Internet or to the digital age, or to an era of environmental crisis, they can take new forms in these times.

## AWARENESS AND CONCERN

A key goal of GlacierHub is to link the first two elements, awareness and concern, in order to expand public engagement with issues of glacier retreat in mountain societies. To promote action, it is not sufficient merely to provide information to a reader of the website. It is also important to evoke in the reader a sense of the significance of that information. However, in a world saturated with information, people who are online are often distracted. The writers at GlacierHub seek to craft material that will draw readers in, by using clear, vivid language, by telling stories with concrete elements and strong narrative development, and by including striking images.

To learn more about the awareness and concerns of the site's readers, the editors and writers at GlacierHub turn to analytics, the statistical information that Google provides on website traffic, and that Facebook, Twitter and Instagram distribute on the social media activity on their respective platforms. These sources indicate how many people have viewed each post, how long they have spent at the post and, in many cases, the location from which they accessed it. They also show which material draws the most attention and comment on social media sites.

The analytics show that the website's audience is international. Table 12.1 presents a summary of these statistics; it distinguishes between large and small countries with glaciers, using a total population of 50 million as the cut-off between the two. (It lists Germany and Spain as minimal glacier countries, since each of those countries have a total glacier area of only a few hundred hectares.) In some cases, such as China and the United States, the large glacier countries have glacier-covered areas that are large in absolute terms, but small in proportion to the country's whole area; the reverse is true for some smaller glacier countries, such as Bhutan and Kyrgyzstan.

The United States accounts for a little over 45 per cent of pageviews, or occasions when a user opens one page within a site. The next two are the UK (8 per cent) and Canada (6 per cent). Among the next ten countries in order of frequency are Switzerland, Peru, Norway and Nepal, and the following ten include New Zealand, Bhutan, Sweden, Iceland and Kyrgyzstan. It is striking to see this representation of a number of glacier countries, including ones with small populations. Bhutan and Iceland both have fewer than 1 million inhabitants, and yet account for more pageviews than much larger countries, indicating a high per capita rate.

Moreover, with the exceptions of Canada, New Zealand and Sweden, these 11 small glacier countries all have a smaller than average percentage of new users, or individuals who are arriving at the site for the first time (some of whom will return, though most will not). Worldwide, 77.69 per cent of users, just over three-quarters, are new users. In glacier countries, a higher proportion of the users return to the website, so the percentage of new users is lower. There is also some suggestion that the visitors from the small glacier countries stay longer on the website than average; their sessions are slightly longer, on average, and they have a higher percentage of sessions over five minutes in length. However, there are no clear differences in another key measure of engagement, the average number of pages per session (a higher number indicates that users go more often from one page to another within an individual session).

This discussion of analytics leads to the debates over 'clicktivism', a term that refers to the use of electronic media by social movements to promote causes and movements and to organise protests. The term appeared in print less than a decade ago. One of the earliest, if not the very first, uses of it occurs in a book *Tactical Media* (Raley 2009), which explores a variety of practices, grouped under the label 'interventionist media art', which disrupt neoliberalism. Though some (Halupka 2014) have followed this line of thought, seeing a great deal of promise in network-based activism, others (Drumbl 2012) find that the simple act of forwarding a post or clicking on a 'like' or 'favourite' often does not lead to other, more effective forms of action. This critique argues that social media can breed facile, shallow engagement with an issue that stifles rather than encourages more sustained action. For this reason, we look for other lines of evidence to assess awareness and concern, in addition to the quantitative measures of traffic at the website.

The analytics at GlacierHub show that readers are interested in posts on scientific aspects of glaciers. Stories that discuss eruptions of glacier-clad volcanoes receive a good deal of traffic. Many factors could contribute to this popularity: the striking contrast of lava and ice, the dramatic photographs which show the eruptions, and the association of the eruptions with natural hazards (ash clouds which interfere with aviation, debris flows that threaten and sometimes damage communities on the slopes of the mountains). Readers have also shown a strong interest in posts on art, whether paintings, photographs or audio projects that link recordings of glacier sounds with instrumental music. And the posts on archaeological finds that have been exposed by retreating ice also draw many viewers, such as a story on a 1,400-year-old sledge that was discovered in Norway (Dietz 2016).

In addition, a number of posts provide information about local adaptation and resilience efforts, including efforts by communities and local non-governmental organisations (NGOs), such as greenhouse construction in Nepal (Dolma 2016) and reuse of materials in abandoned buildings in Switzerland (Miao 2016). Moreover, readers are also drawn to posts with a more explicitly political content. Among the most popular posts are those that link glaciers and mining and that discuss the negative impacts of mining activities on glaciers. One such post

Table 12.1 Use statistics for GlacierHub, first four years

| Nation | Glacier status | Rank | Number of sessions | Percentage of total sessions (%) | Percentage of new users among all visitors (%) | Average number of pages per session | Average length of session in hours, minutes and seconds | Percentage of sessions over 5 minutes in length (%) |
|---|---|---|---|---|---|---|---|---|
| TOTAL | | | 181,712 | 100 | 77.69 | 1.47 | 0:01:04 | 2.94 |
| United States | Large glacier country | 1 | 82,845 | 45.59 | 76.91 | 1.55 | 0:01:19 | 3.16 |
| United Kingdom | None | 2 | 14,673 | 8.07 | 80.91 | 1.32 | 0:00:44 | 2.10 |
| Canada | Small glacier country | 3 | 11,749 | 6.47 | 82.22 | 1.34 | 0:00:54 | 2.71 |
| Russia | Large glacier country | 4 | 8,844 | 4.87 | 66.55 | 1.80 | 0:01:44 | 3.05 |
| India | Large glacier country | 5 | 5,777 | 3.18 | 79.89 | 1.34 | 0:01:11 | 3.79 |
| Australia | None | 6 | 4,847 | 2.67 | 81.53 | 1.33 | 0:00:57 | 2.74 |
| Germany | Minimal glacier country | 7 | 3,758 | 2.07 | 81.37 | 1.42 | 0:00:54 | 2.71 |
| France | Large glacier country | 8 | 3,421 | 1.88 | 80.50 | 1.35 | 0:00:43 | 2.34 |
| Switzerland | Small glacier country | 9 | 2,764 | 1.52 | 64.29 | 1.62 | 0:01:17 | 3.55 |
| Peru | Small glacier country | 10 | 2,442 | 1.34 | 64.33 | 1.44 | 0:01:21 | 4.63 |
| Norway | Small glacier country | 11 | 2,198 | 1.21 | 71.11 | 1.41 | 0:00:56 | 2.82 |
| Nepal | Small glacier country | 12 | 1,987 | 1.09 | 67.24 | 1.38 | 0:01:23 | 3.52 |
| Netherlands | None | 13 | 1,669 | 0.92 | 83.94 | 1.41 | 0:00:58 | 3.00 |
| Brazil | None | 14 | 1,514 | 0.83 | 90.16 | 1.17 | 0:00:34 | 1.19 |

| | | | | | | | |
|---|---|---|---|---|---|---|---|
| Italy | Large glacier country | 15 | 1,454 | 0.80 | 79.64 | 1.43 | 0:01:07 | 3.44 |
| China | Large glacier country | 16 | 1,406 | 0.77 | 84.64 | 1.47 | 0:01:21 | 3.98 |
| New Zealand | Small glacier country | 17 | 1,223 | 0.67 | 83.48 | 1.29 | 0:00:48 | 2.86 |
| Bhutan | Small glacier country | 18 | 1,219 | 0.67 | 73.01 | 1.41 | 0:01:23 | 2.63 |
| Spain | Minimal glacier country | 19 | 1,123 | 0.62 | 86.11 | 1.31 | 0:00:45 | 1.69 |
| Sweden | Small glacier country | 20 | 1,096 | 0.60 | 78.74 | 1.40 | 0:00:51 | 1.92 |
| Austria | Small glacier country | 21 | 1,088 | 0.60 | 57.26 | 2.34 | 0:03:26 | 3.31 |
| Iceland | Small glacier country | 22 | 1,088 | 0.60 | 76.65 | 1.30 | 0:00:52 | 2.57 |
| Kyrgyzstan | Small glacier country | 23 | 1,038 | 0.57 | 50.58 | 1.39 | 0:01:21 | 3.47 |
| Japan | None | 24 | 1,011 | 0.56 | 82.20 | 1.31 | 0:00:47 | 2.67 |
| Chile | Small glacier country | 25 | 981 | 0.54 | 77.06 | 1.51 | 0:01:17 | 3.67 |
| Denmark | None | 26 | 916 | 0.50 | 76.42 | 1.48 | 0:00:55 | 2.40 |
| Poland | None | 27 | 882 | 0.49 | 67.57 | 1.23 | 0:00:40 | 2.49 |
| Singapore | None | 28 | 878 | 0.48 | 91.69 | 1.19 | 0:00:34 | 2.05 |
| Pakistan | Large glacier country | 29 | 776 | 0.43 | 84.28 | 1.27 | 0:01:14 | 3.61 |
| Belgium | None | 30 | 748 | 0.41 | 89.71 | 1.30 | 0:00:38 | 1.87 |

Data source: Google Analytics, September 2014–August 2018.

(French 2015) – one of the five most read posts in the website's history – focused on removing ice from Svartisen Glacier in northern Norway and shipping it to the Persian Gulf, where it is sold to provide high-priced ice cubes as a luxury item. Though glacier ice has some physical properties which could attract customers – it can be presumed to be very pure, it is so dense that it melts slowly, and the air bubbles that are contained within it are so compressed that they make a popping sound when they are released – its appeal seems to rest more on its limited availability, high cost and exotic origin. This story was picked up by other websites, which condemned this mining as wasteful and expressed concerns for possible destabilisation of the glacier itself (Bryce 2015; Pakalolo 2015). These posts attracted comments from readers, who also were disturbed by this unnecessary consumption.

The posts on mining can show the complexity of ecological crises and environmental politics in the contemporary world. One post (Hill 2016) discussed the successful efforts of Native American groups in Washington State in the United States to block the construction of a deep water port that would have exported coal, mined in Wyoming, to Asia. These indigenous communities argued that the port would have disturbed the fishing grounds that were guaranteed to them by treaties in the nineteenth century; the different species of salmon which are protected under the treaties are already threatened by reductions in the flow of water in the rivers where the fish spawn. Glacier retreat, in mountains inland from the coastal site of the proposed port, has been a cause of this reduction, and, as such, is part of the indigenous people's concerns for their fisheries. Through this post, GlacierHub helped publicise this case around the world, and promoted an awareness of the complexity of current ecological crises and energy issues.

Other posts addressed mining efforts in developing countries. Five separate posts discussed the Canadian-owned gold mine, Kumtor, in Kyrgyzstan, discussing the mine's practice of removing glacier ice to reach ore deposits. This removal destabilises the glaciers, and has been the target of protests by local environmental groups (French 2014b; Satke 2015, 2016; Verchot 2015), and, by impacting local hydrology, increases the risk of dam failures (GlacierHub 2015a). Similar removal of glacier ice by mining companies in Chile was also discussed on the website (French 2014a) and reported on Greenpeace and other environmental groups in Chile which oppose such efforts and call for stronger regulation (Sayeed 2016).

In addition to addressing mining, the website has also published several posts on indigenous issues and mountain tourism. In Nepal, tensions have been growing between the Sherpa guides, native to the

Mount Everest region, and the national tourism organisations. In 2014, demonstrations followed after a number of guides died in an accident on Mount Everest, due in part to pressure from the government for the peak to remain open even in times of high risk; the scanty compensation which the families received exacerbated the conflict (GlacierHub 2014). More recently, Sherpa guides have sought more actively to receive official certificates that document their having reached the summit of Mount Everest; these certificates would support their efforts to gain recognition and to have a stronger say in tourism management (Marconi 2016). In Peru, an indigenous community whose traditional lands extend into areas now controlled by the country's largest national park with glaciers had a series of violent confrontations with the park authorities. After long negotiations, they have established a co-management system that gives them responsibility for trash removal and also allows them to provide food, lodging and other services to tourists (Rasmussen 2014; Belew 2018).

Another political theme is risk management, endeavouring to represent the perspectives of high mountain communities to glacier-related hazards. One account from Nepal discusses villagers' fears of floods in the weeks after a major earthquake, indicating that government programmes did not provide sufficient logistical support or recognition of their concerns, derived from their long history in the region (Sherpa 2015). A related post from nearby Bhutan (Orlove 2014b) brings forward interviews with glacier flood survivors to show the cultural dimensions of loss. A post (Fraser 2017) about Peru examines the case of indigenous community members who destroyed the early warning system that Swiss and international NGOs had installed in a flood-prone lake high above the village. It reviews the deficits in the efforts of the NGOs to communicate with the villagers. Fraser suggests that the NGOs relied on technical solutions to risk reduction at the expense of social programmes to incorporate villagers into the monitoring, and points to their cultural insensitivity as well. A more positive case (Davison 2017) from Colombia, at the Nevado del Ruiz, considers a group of filmmakers who worked at length with local villagers to record their memories of earlier glacier-based floods and debris flows and to discuss strategies, such as early evacuations, to promote preparedness and risk reduction. The film was shown at community meetings in other high-risk regions as a means of sparking local participation in hazard planning.

In addition, the website has published posts on glacier-related environmental issues since Donald Trump's election as President of the United States in November 2016. It reported on efforts by Trump's administration to weaken national climate assessments within the United States

(Zarzar 2017). The website issued one of its rare editorials when Trump pulled the United States out of the Paris Agreement (Orlove 2017a), and soon followed this up with a post (Orlove 2017b) that noted the condemnation of this withdrawal in a number of glacier countries, including Iceland, Norway, Switzerland, Peru, Chile, Nepal and New Zealand. The website's first post on Ryan Zinke, the first Secretary of Interior under Trump, indicated that he grew up near Glacier National Park in Montana; it quoted him as citing glacier retreat in the park as evidence that climate change 'is not a hoax', a position at variance with Trump's own (Chappo 2017). The website traced oil and gas leases by the Department of Interior under Zinke which could have polluted glacier regions (Orlove 2018b) and reported on the Department's efforts to halt programmes to protect endangered species in glacier regions (Joshi 2018). Zinke's resignation on 15 December 2018 was due largely to scandals over his personal business interests and over the management of finances in the Department, but it also reflected the pressure from many environmental sources who were concerned by Zinke's efforts to weaken the regulations which protected public lands from excessive exploitation by energy and mining firms (Fears et al. 2018). GlacierHub's posts on mountain issues in the Department added another dimension to public awareness of the Department's environmental policies and practices.

There is some evidence that posts on GlacierHub have helped build awareness of the political issues linked to glacier retreat. The posts on mining in Kyrgyzstan received views from Chile and Peru, and were circulated on social media by Greenpeace Chile, while posts on mining in Chile were viewed from Kyrgyzstan as well as from other mountain countries in Latin America. There are similar patterns of reciprocal interest about natural hazards and about tourism between readers in the Andes and in the high mountains of Asia (Himalayas, Hindu Kush, Karakoram, Tien Shan).

## ACTION AND COMMUNITY

As other cases of engaged anthropology have shown, the movement from awareness to concern can continue, as concern leads to action, and action can build community among individuals and organisations who share concerns and seek to coordinate their actions. The initial step of moving from concern to action often involves simple acts of communication: sharing information with others who are in a similar position and have a similar perspective. These acts of communication can build social ties on which community can be based. Initially, these ties may reside principally on social media platforms, where people follow each other's

Twitter feeds and Facebook pages; later on, these can lead to direct conversations and collaborative projects.

One key form of action for the website has been to publicise events and promote participation in them, such as the World March for Science in April 2017. GlacierHub announced a UNESCO conference more than two months before it was held (Orlove 2015), several days before the organisers began accepting applications. The conference, Resilience in a Time of Uncertainty: Indigenous Peoples and Climate Change, was held at UNESCO headquarters in Paris in November 2015 just before COP21, which led to the Paris Agreement. The conference emphasised issues of indigenous knowledge in developing adaptations to climate change, and sought to increase the participation of indigenous representatives at COP21. The post about the conference circulated actively and drew readers from around the world. It was the means by which some members of high mountain communities, including one speaker at the plenary session, learned of the conference. Most of the participants at the conference remained in Paris for COP21 and continued to talk there. The ties developed at this conference have continued, both informally and through an international NGO, the International Network of Mountain Indigenous Peoples.

In a closely related effort, the website has also promoted public demonstrations. It featured a photo essay (Orlove 2014a), entitled 'How to Make a Glacier in 10 Easy Steps', that showed a simple model glacier – a wheelbarrow filled with crushed ice – that had been taken to the People's Climate March in New York in September 2014. The post included the signs that were prepared to accompany this glacier, and a number of selfies that people at the march had taken with the glacier. This post has been viewed over 2,400 times. Moreover, it seems to be taken seriously, since the mean length of the visit is over three and a half minutes – a long time for a short post.

A second form of action and community-building has been GlacierHub's efforts to build and support social ties between the staff of regional mountain organisations, such as the International Center for Integrated Mountain Development (ICIMOD) in Nepal, the Mountain Societies Research Institute (MSRI) in Kyrgyzstan and the National Institute for Research on Glaciers and Mountain Ecosystems (INAIGEM) in Peru. These organisations, each focused on specific mountain regions (respectively, the Himalayas and Hindu Kush; the Tien Shan and Pamirs; the Andes) have broadly similar mandates to conduct research in the natural and social sciences to promote sustainable and participatory livelihoods in high mountain regions, addressing issues of development, equity and disaster risk reduction in the context of climate change. The activities of these organisations in addressing glacier lake outburst floods have attracted particular attention. These organisations have some ties through international mountain organisations such as Mountain Forum, a unit within the United Nations Food and Agriculture Organization, and two international NGOs, the Mountain Institute and the Mountain Research Initiative. These organisations often operate independently, however, and are not in regular communication with each other, but the website provides a conduit for them to stay apprised of each other's activities, and to establish contacts with one another.

The website has also covered international efforts to link mountain communities and nations worldwide. A side meeting at COP21 in Paris in 2015 was discussed on the website, attended by representatives of seven glacier countries, Tajikistan, Bhutan, Peru, Bolivia, Switzerland, Austria and Norway, from three continents (GlacierHub 2015b). And the website covers international climate negotiations from a glacier-centred perspective. One post (Orlove 2016) noted that the three glacier countries, Austria, Bolivia and Nepal, were among the ten which ratified the Paris Agreement on 5 October 2016, putting the total number of supporters above the threshold required for the Agreement to enter into force.

## DISCUSSION AND CONCLUSION

This review suggests that GlacierHub's publications, and the actions and social ties which it has promoted, correspond, at least in part, to some of the avenues that Low and Merry proposed for engaged anthropology. The website is unambiguously a form of connection with media, and a kind of witnessing of violence and social change. The presence of pieces written by community members, activists and journalists from glacier countries points towards co-production of knowledge; the linking with activities, ranging from conferences to protests, can be seen as addressing social problems. Creating solutions, the final avenue that Low and Merry list, is a crucial one. GlacierHub's efforts to spread the word about community efforts, whether in hazard preparedness in Colombia, village greenhouses in Nepal, indigenous conferences in France or landscape restoration in Switzerland, could be seen as promoting solutions.

Of Low and Merry's forms of engagement, GlacierHub has worked consistently towards collaboration, both by publishing pieces from small glacier countries such as Nepal and Kyrgyzstan, and by developing and supporting an international readership which sustains an interest in the issues of high mountain areas worldwide, addressing the global challenge of climate change. In its support of indigenous movements and organisations, it can be seen as engaged in advocacy and activism as well, though it seeks to remain within the conventions of journalism, with a clear distinction between the many articles of objective reporting and the occasional opinion pieces, listed as editorials. This emphasis on journalism supports the inclusion of many posts on science and art, and, to judge from the comments on social media such as Twitter and Facebook, is part of the website's appeal to its readers. In these ways, GlacierHub serves as an example of media-based engaged anthropology.

To date, the website has remained relatively small, while the need to address the ecological crises in high mountain areas, compounded by inequalities within and between nations, is large. Nonetheless, the website has contributed directly to some efforts. The website has built awareness of a Special Report on Oceans and the Cryosphere, being prepared by the Intergovernmental Panel on Climate Change (IPCC). This report, scheduled for completion in 2019, links the loss of ice, in mountain glaciers and in the Antarctic and Greenland Ice Sheets, to sea level rise, and attends to a host of other transformations in high mountain regions, coasts, coastal water and the deep ocean. This special report represents an opportunity for the concerns of high mountain communities to be brought together with the Arctic peoples and small island states that are also within the scope of this report. GlacierHub has contributed to public awareness of the report by reporting on the meetings of the lead authors of the report, and by distributing calls for representatives of mountain organisations and communities to participate as expert reviewers of drafts of the report (Orlove 2017c, 2017d, 2018a, 2018c, 2018d). Other opportunities may arise as well, through other linkages around issues of sustainable development and indigenous rights, signalled through the United Nations' Sustainable Development Goals and through other international forums. The website helped support the UN Declaration on the Rights of Peasants and Other People Working in Rural Areas, which makes explicit reference to the rights of mountain peoples, by tracing its path through UN bodies, particularly the United Nations Human Rights Council, to its approval by the UN General Assembly on 17 December 2018; in this way, it participated in the groups of organisations that built awareness and support for this declaration (Orlove 2018e). As with other efforts of engaged anthropology, the work of GlacierHub will not be carried out by the website alone, but through the networks and coalitions in which it participates and to which it contributes.

## ACKNOWLEDGEMENTS

Major funding for the website and research was provided under the cooperative agreements NSF SES-0345840 and NSF SES-0951516 awarded to the Center for Research on Environmental Decisions at Columbia University in New York. We are grateful for helpful comments from Thomas Hylland Eriksen, Astrid Stensrud and the anonymous reviewers, and for Aiyana Lynn Bodi's assistance in the preparation of the manuscript.

REFERENCES

Websites last accessed 22 February 2019.

Belew, N. 2018. Adapting to Glacier Retreat in Peru's Huascarán National Park. GlacierHub, 8 August. Available at: http://glacierhub.org/2018/08/08/glacial-retreat-in-perus-huascaran-national-park/.

Bryce, E. 2015. This Company Thinks You'll Pay for Luxury Ice from a Shrinking Glacier. Modern Notion, 2 June. Available at: http://modernnotion.com/company-thinks-youll-pay-luxury-ice-shrinking-glacier/.

Chappo, A. 2017. BREAKING: Ryan Zinke Confirmed as Interior Secretary, Talks Glacier Retreat. GlacierHub, 1 March. Available at: http://glacierhub.org/2017/03/01/zinke-trumps-pick-for-interior-talks-glacier-retreat/.

Checker, M. 2009. Anthropology in the Public Sphere, 2008: Emerging Trends and Significant Impacts. *American Anthropologist* 111(2): 162–9.

Davison, H. 2017. Using Film to Reduce Risk on Volcanoes. GlacierHub, 1 June. Available at: http://glacierhub.org/2017/06/01/using-film-to-reduce-risk-on-glacier-capped-and-non-glacier-capped-volcanoes/.

Dietz, G. 2016. Prehistoric Sledge Found on Norwegian Glacier. GlacierHub, 21 June. Available at: http://glacierhub.org/2016/06/21/1400-year-old-sledge-thawed-out-of-norwegian-glacier/.

Dolma, T. 2016. Greenhouses Bring Hope to Vulnerable Mountain Communities in Nepal. GlacierHub, 8 September. Available at: http://glacierhub.org/2016/09/08/a-simple-greenhouse-brings-hope-to-vulnerable-mountain-communities-in-nepal/.

Drumbl, M.A. 2012. Child Soldiers and Clicktivism: Justice, Myths, and Prevention. *Journal of Human Rights Practice* 4(3): 481–5.

Fears, D., J. Eilperin and J. Dawsey. 2018. Zinke was a Rising Star in Washington. Then He Joined the Trump Administration. *Washington Post*, 20 December. Available at: www.washingtonpost.com/energy-environment/2018/12/20/zinke-was-rising-star-washington-then-he-joined-trump-administration/?noredirect=on&utm_term=.f9bff3fc7f6b.

Fraser, B. 2017. Learning from a Flood-alarm System's Fate. GlacierHub, 31 May. Available at: http://glacierhub.org/2017/05/31/learning-from-a-flood-alarm-systems-fate/.

French, K. 2014a. Copper Versus Ice: Chilean Mine Would Excavate Five Glaciers. GlacierHub, 1 October. Available at: http://glacierhub.org/2014/10/01/copper-versus-ice-chilean-mine-would-excavate-five-glaciers/.

—— 2014b. Kumtor Gold Mine Threatens Central Asian Glaciers and Water. GlacierHub, 23 December. Available at: http://glacierhub.org/2014/12/23/kumtor-gold-mine-threatens-central-asian-glaciers-and-water/.

—— 2015. Mining a Norwegian Glacier for Luxury Ice Cubes. GlacierHub, 26 February. Available at: http://glacierhub.org/2015/02/26/mining-a-norwegian-glacier-for-luxury-ice-cubes/.

Gertner, J. 2009. Why Isn't the Brain Green? *New York Times Magazine*, April. Available at: www.nytimes.com/2009/04/19/magazine/19Science-t.html.

GlacierHub. 2014. In Wake of Everest Deaths, Many Groups Push for Reform. GlacierHub, 16 September. Available at: http://glacierhub.org/2014/09/16/in-wake-of-everest-deaths-many-groups-push-for-reform/.

—— 2015a. Dam Spill Threats at a Gold Mine in Kyrgyzstan. GlacierHub, 17 February. Available at: http://glacierhub.org/2015/02/17/a-risk-of-a-dam-spill-threatens-a-large-gold-mine-in-the-mountains-of-kyrgyzstan/.

—— 2015b. Meeting at COP21 Seeks Coordination of Glacier Countries. GlacierHub, 10 December. Available at: http://glacierhub.org/2015/12/10/international-meeting-at-cop21-seeks-coordination-of-glacier-countries/.

—— 2018. About GlacierHub. Available at: http://glacierhub.org/about/.

Goldstein, D.M. 2014. Laying the Body on the Line: Activist Anthropology and the Deportation of the Undocumented. *American Anthropologist* 116(4): 839–42.

Habermas, J. 1989. *The Structural Transformation of the Public Sphere: An Inquiry into a Category of Bourgeois Society*. Cambridge: Polity Press.

Halupka, M. 2014. Clicktivism: A Systematic Heuristic. *Policy & Internet* 6(2): 115–32.

Hill, S. 2016. Washington State Tribes Block Coal Terminal. GlacierHub, 30 June. Availabe at: http://glacierhub.org/2016/06/30/washington-state-tribes-vulnerable-to-climate-change-despite-coal-terminal-block/.

Kuklick, H. 2009. *A New History of Anthropology*. London: Wiley-Blackwell.

Joshi, S. 2018. Restoration of Grizzly Bear Population in North Cascades Halted. GlacierHub, 30 January. Available at: http://glacierhub.org/2018/01/30/restoration-of-grizzly-bear-population-in-north-cascades-halted/.

Lamphere, L. 2004. The Convergence of Applied, Practicing, and Public Anthropology in the 21st Century. *Human Organization* 63(4): 431–43.

Low, S.M. and S.E. Merry. 2010. Engaged Anthropology: Diversity and Dilemmas: An Introduction to Supplement 2. *Current Anthropology* 51(S2): S203–S226.

Marconi, B. 2016. High Altitude Himalayan Heroes Denied Summit Certificates. GlacierHub, 19 October. Available at: http://glacierhub.org/2016/10/19/high-altitude-himalayan-heroes-denied-summit-certificates/.

Miao, X. 2016. In an Empty Building's Place: Wilderness and Community. GlacierHub, 14 January. Available at: http://glacierhub.org/2016/01/14/in-an-empty-buildings-place-wilderness-and-community-2/.

Mullins, P.R. 2011. Practicing Anthropology and the Politics of Engagement: 2010 Year in Review. *American Anthropologist* 113(2): 235–45.

Orlove, B. 2014a. How to Make a Glacier in 10 Easy Steps. GlacierHub, 23 September. Available at: http://glacierhub.org/2014/09/23/how-to-make-a-glacier-in-10-easy-steps/.

—— 2014b. Bhutan's Fortresses Yet Another Victim of Glacial Floods. GlacierHub, 11 November. Available at: http://glacierhub.org/2014/11/11/bhutans-fortresses-yet-another-victim-of-glacial-floods/.

—— 2015. UNESCO Conference on Indigenous Peoples and Climate. GlacierHub, 1 September. Available at: http://glacierhub.org/2015/09/01/unesco-conference-indigenous-peoples-climate-change/.

—— 2016. Glacier Countries Help the Paris Agreement Enter into Force. GlacierHub, 6 October. Available at: http://glacierhub.org/2016/10/06/glacier-countries-help-the-paris-agreement-enter-into-force/.

—— 2017a. Cracks in the Paris Agreement. GlacierHub, 1 June. Available at: http://glacierhub.org/2017/06/01/cracks-in-the-paris-agreement/.

—— 2017b. Glacier Countries Condemn Trump's Withdrawal from the Paris Agreement. GlacierHub, 19 June. Available at: http://glacierhub.org/2017/06/19/glacier-countries-condemn-trumps-withdrawal-from-the-paris-agreement/.

—— 2017c. IPCC Announces Details of a Report Chapter on High Mountains. GlacierHub, 23 August. Available at: http://glacierhub.org/2017/08/23/ipcc-announces-details-of-a-report-chapter-on-high-mountains/.

—— 2017d. Glacier Researchers Gather at IPCC Meeting in Fiji. GlacierHub, 19 October. Available at: http://glacierhub.org/2017/10/19/glacier-researchers-gather-ipcc-meeting-fiji/.

—— 2018a. Glacier Researchers Gather at IPCC Meeting in Ecuador. GlacierHub, 27 February. Available at: http://glacierhub.org/2018/02/27/glacier-researchers-gather-ipcc-meeting-ecuador/.

—— 2018b. Is the Department of the Interior Taking Steps to Protect Montana's Glaciers? GlacierHub, 15 March. Available at: http://glacierhub.org/2018/03/15/department-interior-taking-steps-protect-montanas-glaciers/.

—— 2018c. IPCC Report is Now Open for Comment. GlacierHub, 8 May. Available at: http://glacierhub.org/2018/05/08/ipcc-report-is-now-open-for-comment/.

—— 2018d. Glacier Researchers Gather at IPCC Meeting in China. GlacierHub, 2 August. Available at: http://glacierhub.org/2018/08/02/glacier-researchers-gather-at-ipcc-meeting-in-china/.

—— 2018e. UN Resolution Recognizes the Rights of Rural Peoples. GlacierHub, 4 October. Available at: http://glacierhub.org/2018/10/04/un-resolution-recognizes-the-rights-of-rural-peoples/.

Orlove, B., H. Lazrus, G.K. Hovelsrud and A. Giannini. 2014. Recognitions and Responsibilities: On the Origins and Consequences of the Uneven Attention to Climate Change Around the World. *Current Anthropology* 55(3): 249–75.

Osterweil, M. 2013. Rethinking Public Anthropology Through Epistemic Politics and Theoretical Practice. *Cultural Anthropology* 28(4): 598–620. https://doi.org/10.1111/cuan.12029.

Pakalolo. 2015. The Super Rich Will Soon Have Their Russo-Baltique Vodka Martini Chilled By Thousand Year Old Ice. *Daily Kos*, 11 March. Available at: www.dailykos.com/stories/2015/3/11/1367499/-The-super-rich-will-soon-have-their-Russo-Baltique-Vodka-martini-chilled-by-thousand-year-old-ice.

Raley, R. 2009. *Tactical Media*. Minneapolis, MN: University of Minnesota Press.
Rasmussen, B.M. 2014. Traces of Tourism at the Peru Glacier are More than Footprints. GlacierHub, 10 September. Available at: http://glacierhub.org/2014/09/10/traces-of-tourism-at-the-pastoruri-glacier-leaves-more-than-footprints/.
Satke, R. 2015. European Bank Says Mining Projects Don't Damage Glaciers. GlacierHub, 8 July. Available at: http://glacierhub.org/2015/07/08/european-bank-says-mining-projects-dont-damage-glaciers/.
—— 2016. Mining Company Shirks Blame for Glacier Damage in Kyrgyzstan. GlacierHub, 20 April. Available at: http://glacierhub.org/2016/04/20/mining-company-shirks-blame-for-glacier-damage-in-kyrgyzstan/.
Sayeed, S. 2016. Activists Say Chilean Glacier Protection Law Falls Short. GlacierHub, 5 May. Available at: http://glacierhub.org/2016/05/05/activists-say-chilean-glacier-protection-law-falls-short/.
Sherpa, P.Y. 2015. Nepali Villagers Trapped Under Threat of Glacier Floods. GlacierHub, 28 May. Available at: http://glacierhub.org/2015/05/28/nepali-villagers-trapped-threat-glacier-floods/.
Verchot, M. 2015. No Mines in Kyrgyzstan Exacerbate Glacier Advance. GlacierHub, 20 August. Available at: http://glacierhub.org/2015/08/20/mines-in-kyrgyzstan-exacerbate-glacier-advance/.
Zarzar, R. 2017. National Climate Assessment Report Under Review by Trump Administration. GlacierHub, 22 August. Available at: http://glacierhub.org/2017/08/22/national-climate-assessment-report-rejected-by-the-trump-administration/.

# Notes on Contributors

**Astrid Oberborbeck Andersen** holds a PhD in Social Anthropology from the University of Copenhagen. She is Assistant Professor in Techno-Anthropology at the Department of Learning and Philosophy, Aalborg University. Her research interests include human-environment relations, interfaces between knowledge and politics, and the roles technology plays in these; societal aspects of climatic changes and natural resources; and interdisciplinarity. She has carried out research in Peru and Northwest Greenland.

**Janne Flora** holds a PhD from the Scott Polar Research Institute, University of Cambridge. She is currently a postdoc at the Department of Bioscience, Aarhus University, and an affiliated researcher at the Stefansson Arctic Institute in Akureyri. She is the author of 'Wandering Spirits: Loneliness and Longing in Greenland' (Chicago).

**María A. Guzmán-Gallegos** is a postdoctoral fellow at the University of Oslo, and has done extensive fieldwork in Kichwa communities in Ecuadorian and Peruvian Amazonia. Her research focus is on socio-environmental conflicts related to extractive industries, indigenous movements, Amerindian conceptualisations of personhood, socio-natural worlds, interethnic relations and socio-environmental conflict related to extraction.

**Kirsten Hastrup** is Professor of Anthropology at the University of Copenhagen. Over the past ten years she has worked on Greenlandic issues, mainly related to the Thule region where she has carried out recurrent fieldwork. Before that, her main field was Iceland. She has published numerous articles and books on both of these fields, with a sustained focus on the interface between nature and society, both historically and contemporary.

**Edvard Hviding** is Professor of Social Anthropology at the University of Bergen, founding director of the Bergen Pacific Studies Research Group and director of the SDG Bergen Task Force. He is Principal Investigator of the project '*Mare Nullius*? Sea level rise and maritime sovereignties in the Pacific – an expanded anthropology of climate change'. His research in the Pacific has been ongoing since 1986, with more than five years of field research mainly in the Solomon Islands. He is currently engaged in high-level science advice for ocean and climate diplomacy of the Pacific Islands states and organisations, at the United Nations and regionally.

**Thomas Hylland Eriksen** is Professor at the Department of Social Anthropology, University of Oslo. His interest in globalisation has led his research from studies of ethnic relations, nationalism and the post-ethnic to research on accelerated change, technology and the environment. His books with Pluto Press include *Small Places, Large Issues: An Introduction to Social and Cultural Anthropology* (fourth edition 2015); *Ethnicity and Nationalism: Anthropological Perspectives* (third edition 2010); *A History of Anthropology* (second edition 2013); *Overheating: An Anthropology of Accelerated Change* (2016); and *Boomtown: Runaway Globalisation on the Queensland Coast* (2018).

**Andrei Marin** is a human geographer and researcher at the Department of International Environment and Development Studies, Norwegian University of Life Sciences. His work investigates the interface between environmental and social change, with particular focus on climate change vulnerability and adaptation and institutional change. He has worked with these issues in Mongolia, Norway and Kenya. The research presented herein derives in part from his latest project (2015–18) 'Socio-ecological transformations: human animal relations under climate change in Northern Eurasia', funded by the Research Council of Norway (project number 244907/E10).

**Kerry Milch** is the Assistant Director for Undergraduate Enrichment at Temple University in Philadelphia, where she works to broaden undergraduate participation in research and creative endeavours. She holds a PhD from Columbia University in Psychology and worked for several years as an Associate Research Scholar at Columbia's Center for Research on Environmental Decisions (CRED). During her time at CRED, Kerry studied topics such as perceptions of climate change among glacier communities and why people fail adequately to prepare for hurricanes. She also worked with the US Geological Survey to create a natural hazards communication guide based on insights from social science research.

**Mark Nuttall** is Professor and Henry Marshall Tory Chair in the Department of Anthropology at the University of Alberta. He is also visiting Professor of Climate and Society at the University of Greenland and Greenland Climate Research Centre in Nuuk. He has carried out extensive research in Greenland, Canada, Alaska, Finland and Scotland and is author and editor of several books, including *Climate, Society and Subsurface Politics in Greenland: Under the Great Ice* (2017); *The Scramble for the Poles: The Geopolitics of the Arctic and Antarctic* (2016, co-authored with Klaus Dodds); *Anthropology and Climate Change: From Actions to Transformations* (2016, co-edited with Susan Crate); and *The Routledge Handbook of the Polar Regions* (2018, co-edited with Torben Christensen and Martin Siegert). He is Fellow of the Royal Society of Canada and a member of the Norwegian Scientific Academy for Polar Research.

**Ben Orlove** is Professor of International and Public Affairs at Columbia University, where he is also Director of the Center for Research on Environmental Decisions (CRED), Director of the Master's Programme in Climate and Society and Senior Research Scientist at the International Institute for Climate and Society. He holds a BA in Anthropology from Harvard and an MA and PhD in Anthropology from the University of California at Berkeley. He is a Fellow of the American Association for the Advancement of Science, and a recipient of a Presidential Award from the American Anthropological Association. He works broadly on social, economic, environmental and policy issues.

**Cecilia G. Salinas** has a PhD in Social Anthropology from the University of Oslo. She is currently Head of Studies and Lecturer at the Department of International Studies and Interpreting at the Oslo Metropolitan University. Her main areas of research are nature conservation, indigenous people, policy and politics. She has published a book on the pulp industry and social movements in Uruguay (*Añorada Esperanza: Respuestas Locales a las políticas neoliberales. Uruguay y la industria de la celulosa*, 2011) and articles on the UN REDD programme, UNESCO biosphere reserve, painting and anthropology and on corporate social responsibility.

**Frank Sejersen** has an MA in Anthropology and a PhD from the Faculty of Humanities, both from the University of Copenhagen. He is Associate Professor at the Department of Cross-Cultural and Regional Studies, University of Copenhagen. Since the end of the 1980s, he has been working with the human dimension of resource management, indigenous peoples' politics, economics, urbanisation, climate change and issues of self-determination in the Arctic. Lately, his analytical attention has been on the politics of hope, conflicting temporalities, the politics of scaling and the emergence of resources in processes related to extractive industries. He has pursued fieldwork in the Greenlandic towns of Sisimiut and Nuuk.

**Astrid B. Stensrud** is Assistant Professor in Development Studies at the University of Agder. She holds a PhD in Social Anthropology from the University of Oslo. Until 2018, she was a postdoctoral fellow at the

Department of Social Anthropology, University of Oslo, where she was part of the 'Overheating' research project. She has done extensive ethnographic fieldwork in Peru and has published articles and chapters on climate change, neoliberalism and water politics in various journals and edited books.

**Anna Tsing** is Professor of Anthropology at the University of California, Santa Cruz, and a Niels Bohr Professor at Aarhus University, where she co-directs Aarhus University Research on the Anthropocene. Her latest work concerns a digital environmental studies project, 'Feral atlas: the more-than-human Anthropocene'.

**Laura Uguccioni** is a graduate of the Master of Arts programme in Quantitative Methods in the Social Sciences at Columbia University, and holds a Master of Food and Resource Economics from the University of British Columbia and a Bachelor of Arts in Economics and Philosophy from the University of California at Santa Barbara. Her professional experience includes data analytics at the Urban Design Lab at Columbia University, and research at Sampark, a non-governmental organisation providing policy advisory services and economic development interventions. Her research interests are economic policy, text analytics and statistics.

**Harold Wilhite** is Professor Emeritus of Social Anthropology at the University of Oslo's Centre for Development and Environment. He has published widely on the theories, practices and policies of consumption, energy savings and sustainable development, drawing on ethnographic fieldwork in several parts of the world, including Central America, USA, Norway, Japan and India. His books include the *Consumption and the Transformation of Everyday Life: A View from South India* (2008) and *Political Economy of Low Carbon Transformation: Breaking the Habits of Capitalism* (2016).

# Index

Abraham, Itty, 25–6
accumulation, 8, 9, 11, 14, 26, 124, 140, 163, 164, 199
Acosta, A., 153, 154
adaptation, 2, 5, 15, 60, 76, 78, 80, 98, 106–8, 117, 127–9, 130, 208, 217, 225
Africa, 33, 35, 152, 153, 157, 158, 161, 206
agrarian reform, 33, 98, 99
agribusiness, 13, 100, 104, 106, 152, 153
agriculture, 1, 31, 76, 96, 97, 98, 99, 102, 107, 108, 109, 153, 161, 175
Ahlgren, Ingrid, 180, 184
AIDESEP (Interethnic Association for the Development of the Peruvian Rainforest), 162
Alaska, 188, 190, 192
  Kivalina, 188, 190–4, 202
Alcoa, 196, 197, 198, 199
algal blooms, 32, 34, 35
Alvarez, José, 133, 140
Amazonia, 26, 29, 162
  Peru, 14, 133–50, 162
  rainforest, 13
  resources, 13, 162
  rivers, 3, 14
Americas, 45, 153, 158
ammonia, 29–30, 37–8n15, 141
Andersen, Astrid Oberborbeck, 6, 12
Andes mountains, 3, 13, 96–114, 216, 217
Antarctic, 13, 219
Anthropocene, 6–10, 22, 24, 25, 32, 37n8, 52, 179
anthropology, 3, 4, 4–6, 8, 10, 16, 17, 119, 129, 130, 173, 205–23

Arctic, 2, 42, 43, 44, 57, 59, 115–32, 188, 201
  High Arctic Thule Region in Northwest Greenland, 8, 41, 48, 49
  indigenous people, 60, 207, 219
  militarisation of, 8
  natural resources, 13, 50, 57–8
  New Arctic, 41, 42
  overheating, 54, 57, 58, 59, 192, 196
  polar bears, 70
  Thule, 8, 41–56, 118, 119, 127
Argentina, 159, 162
Asdal, K., 136, 140, 145
Asia, 33, 34, 59, 152, 153, 178, 214, 216
Atlantic Ocean, 15, 45, 59, 97
Augustinian Catholic Mission, 133, 140
Australia, 14, 136, 145, 172, 182, 183, 209
Austria, 218

Bali, 160
Barents Sea, 59
Barnes, Jessica, 3
Bartra Producción, 133, 138
Bateson, Gregory, 6, 128
Bering Sea, 59
Bhutan, 210, 215, 218
Bikini Atoll, 7, 25
biodiversity, 10, 133, 159, 164, 165, 174
birds, 34, 51, 127, 144, 182
Bolivia, 106, 154, 209, 218
Bosch, Carl, 29–30
Brazil, 25, 26, 29, 159
Brightman, M., 9
Brown, Kate, 27

Brussels, 173
Bryant, Rebecca, 189, 190

California, 77
Canada, 45, 69, 77, 134, 172, 210, 214
carbon dioxide ($CO_2$), 1, 2, 4, 7, 22, 35, 57, 154, 159, 160
carbon sinks, 151, 159, 165
carbon trade, 12, 161-2, 165
Carson, Rachel, 31, 34
cashmere, 13, 76, *77 fig. 5.2*, 79-82, 83, 84-5, 86-7, 87-8, 89
CGIAR (Consultative Group on International Agricultural Research), 33, 35
Chakrabarty, Dipesh, 14
Chile, 214, 216
China, 33, 209, 210
  cashmere, 13, 76, *77 fig. 5.2*, 80, 82, 83, 84, 85, 87
  eBay China, 119, 120, 128
  emissions, 159
  fishing, 183
  forests, 154
Chukchi Ocean, 191
CITES (conventions on the protection of endangered species), 125, 130n5
Clean Development Mechanism (CDM), 151, 160, 163
climate change denial, 2, 5, 36
climate justice, 16, 173
coal, 1, 8, 85, 154, 159, 165, 214
Cold War, 3, 7, 8, 26-40, 42, 46-50, 54, 85, 206
Colombia, 215, 218
colonialism, 15, 42
  Africa, 206
  global South, 151, 153, 165
  Greenland, 41, 54, 194, 197, 198, 202
  India, 154, 155-9
  Latin America, 97, 109, 110
  Oceania, 178, 181, 183
  *terra nullius*, 172
Columbian Exchange, 15
Columbia University, 182, 208-9, 220

commodification, 5, 8, 11, 120, 127, 152
communism, 28, 33, 35, 79, 86, 88
Cook Islands, 183
COP21 (2015), 2, 217, 218
coral reefs, 14, 171, 172, 173, 175, 176, 178, 179, 182, 184
Crate, Susan, 13

DDT, 30-31, 34
dead zones, 32, 35
deforestation, 152, 153, 156, 160, 161
Denmark, 42, 46, 47, 58, 65, 115, 194, 197, 201
deregulation, 4, 88, 97, 98-100, 103, 109, 134
de Soto, Hernando, 98-9
DIGESA (National Environmental Health Agency), 141, 142
displacement, 1, 26, 137, 138, 156, 157, 171, 173, 179, 181, 182
Distant Early Warning (DEW) Line radar stations, 8, 48, 50
Douglas, M., 16
droughts, 1, 77, 78, 80, 96, 101, 102, 103, 171, 178, 179

Earth Summit in Rio de Janeiro (1992), 26, 159
Ecuador, 106, 136, 137
EEZs (Exclusive Economic Zones), 13, 173, 176, 182, 183, 184
Einstein, Albert, 14
Eisenhower, Dwight D., 33
emissions, 16, 166
  carbon, 1, 2, 4, 5, 7, 152, 154, 159, 160, 162, 166
  greenhouse gas, 5, 6, 178, 182, 192, 203, 211
Enlightenment, 42, 157
Escobar, A., 97, 163, 164
Europe, 42-3, 59, 64, 97, 104, 134, 136, 145, 172, 176, 178
  Eastern Europe, 79
  European Union, 64, 183
  Western Europe, 15

Everest, Mount, 215
extinction, 1, 8, 14, 22, 29, 32, 35, 41–56, 111, 154
extractive industries, 8, 13, 50, 54, 58, 68, 69, 106, 151–4, 164
extractivism, 9, 108, 153

Facebook, 118, 119, 120, 129, 210, 217, 219
fair trade, 13, 88–90
farming, 13, 31, 42–3, 78, 79, 101, 137, 156, 158, 159, 161, 162, 165
  farmers, 32, 33, 77, 90, 96–114
Fiji, 173, 181
First World War, 30
fishing, 3, 12, 13, 34, 45, 59, 61–6, 67, 68, 73, 115–32, 138, 142, 173, 175, 182–4, 194, 195, 196, 114
flooding, 1, 171, 173, 178, 179, 180, 215, 218
Ford, Gerald, 32
Ford Foundation, 33
forests, 10, 11, 13, 32, 134, 136, 137, 139, 143, 146, 151–70
  rainforests, 13, 26, 153, 162, 175
fossil fuels, 2, 3, 7, 8, 12, 36
Foster, J.B., 9, 10
Foucault, Michel, 88
France, 209, 218

gas, 1, 2, 58, 59, 69, 71, 159, 160, 192, 216
Germany, 210
GlacierHub, 16, 205–23
glaciers, 16, 22, 48, 50, 57, 59, 61, 62, 68, 71, 101, 102, 196, 205–23
Global Footprint Network, 5
globalisation, 4, 12, 15, 16, 76–94, 109, 115–32, 188, 200, 201
global North, 27, 29, 58, 110, 153
global South, 10, 15, 27, 29, 33, 151–70
global warming, 1, 13–14, 41, 66, 101, 123, 160, 192, 194
Google, 210
Gore, Al, 85

Great Acceleration, 6–10, 22–40, *23 fig. 2.1*
green capitalism, 5, 10, 163–4
greenhouse gas, 5, 6, 178, 182, 192, 203, 211
Greenland, 36, 8, 12, 14, 115–32, 188, 194, 195–204, 219
  Baffin Bay, 45, 59, 68, 69
  Camp Century, 49, 50
  Maniitsoq, 188, 194–201, 202
  Melville Bay, 43, 57, 58, 59, 68, 69, 70, 71
  Northwest Greenland, 41–56, 57–75
  Nuuk, 65, 68, 70, 118
  Qaanaaq, 48, 51, 70, 115–32
Green Revolution, 32, 33, 34, 35
Guzmán-Gallegos, María A., 6, 12, 13, 14

Haber, Fritz, 29–30
Harvey, David, 11, 14, 15, 26, 98
Hastrup, Kirsten, 4, 8, 97, 128
Hau'ofa, Epeli, 172, 177
Hayes, Isaac I., 44–5
herders, 13, 76–95, *77 fig. 5.2*
Himalayas, 13, 216, 217
Hornborg, Alf, 12, 15
hunting, 8, 12, 44–56, 57–65, 115–16, 119–29, 138, 139, 152, 158, 191, 195
hydrocarbons, 58, 68, 69, 70, 148, 140, 141, 144, 145
hydroelectric energy, 157, 196, 197

Iceland, 209, 210, 216, 224
IMF (International Monetary Fund), 78, 79, 83, 85
India, 154, 155–9, 209
  Chipco movement (1973), 156
indigenous people, 16, 27, 35, 60, 134, 152, 158, 161, 162, 177, 183, 188, 194, 214, 215, 217
  organisation, 26, 57, 141, 144, 156, 159, 202, 207
Indochina, 42

Indonesia, 154, 183
inequality, 1, 4, 5, 13–17, 24, 84, 97, 109, 120, 164, 188, 196, 198, 209
Instagram, 210
Intergovernmental Panel on Climate Change (IPCC), 2, 6, 15, 58, 160, 219
Inuit people, 62, 192, 194
Italy, 26

Japan, 25, 32, 33, 83, 87, 180, 183

Kane, Elisha K., 43, 44, 45
Keynes, John Maynard, 9
Kiribati, 172, 173, 174, 175, 176, 180, 181, 184
Klein, Naomi, 153
Korea, 26, 33, 183
Kyoto Protocol, 159
Kyrgyzstan, 209, 210, 214, 216, 217, 218

Latour, Bruno, 8
Liberia, 162
livestock, 29, 76, 79, 80, 81, 153, 158, 161
logging, 153, 156, 157, 161
Low, S.M., 207, 208

Malaurie, Jean, 47, 48
Marin, Andrei, 13, 99, 100
Marshall Islands, 174, 175, 176, 180, 181, 182, 184
Marx, Karl, 9
Mead, Margaret, 206
media, 58, 89, 109, 126, 143, 156, 157, 192, 207, 211, 218, 219
Melanesia, 173, 176
Merry, S.E., 207, 208
Mexico, 33, 77, 161
Micronesia, 175, 176, 184
Mignolo, W.D., 97, 109
mining, 13, 36n1, 53, 54, 100, 106, 107, 153, 157, 184, 214
    coal, 154
    deforestation, 156, 161
    glaciers, 16, 208, 211–14, 216
    Greenland, 121, 122, 200
    mining companies, 58, 68, 85
    Mongolia, 76, 87
    Peru, 134, 135, 141, 143
mitigation, 2, 10, 15, 106, 151–70
modernisation, 10, 27, 32, 36, 48, 98, 106, 107, 108, 109, 110
Mongolia, 3, 13, 76–95
Montana, 216
Mozambique, 158
Muscle Shoals, 30

narwhal, 12, 45, 51, 52, 59, 64, 65, 66, 70–71, 116, 119, 121–5, 127
NASA, 2, 59
National Oceanic and Atmospheric Administration (NOAA), 2
Nauru, 175
Nelson, D., 135, 142, 143
neoliberalism, 4, 11, 78, 82, 84, 85, 88, 89, 90, 91, 96–114, 146, 164, 196, 211
Nepal, 86, 209, 210, 211, 214, 215, 216, 217, 218
New Guinea, 161
New Mexico, 22, 25
New York, 173, 217, 220
New Zealand, 174, 183, 210, 216
Niue, 175
Norway, 2, 136, 160, 210, 211, 214, 216, 218, 225
Oslo, 29
nuclear reactors, 7, 8, 25, 49
Nuttall, Mark, 13, 125

Oceania, 171–87
OEFA (Agency for Environmental Assessment and Control), 141, 142
oil, 1, 69, 80, 121, 123, 133–50, 159, 191, 192, 200, 216
    companies, 14, 70, 71–2, 140
    exploration, 58, 65, 69, 70, 71, 139
    extraction of, 2, 6, 68, 133–7, 143, 146

pipelines, 137, 138
reserves, 13, 137
spills, 3, 14, 127, 133, 138, 139, 140, 141, 144, 146–8
Orlove, Ben, 16
OSINERGMIN (Regulatory Body for Energy Investment and Mining), 141, 142
OXY (Occidental Petroleum), 133, 134, 136, 137, 139, 140

Pacific islands, 13, 30, 171–87
Pacific Ocean, 3, 13, 30, 43, 59, 171–87, 206
Papua New Guinea, 11, 175
Palau, 175
Paris, 173, 217
  COP21, 2, 218
Paris Agreement, 216, 217, 218
pastoralism, 3, 35, 76–95, 107, 158
People's Climate March in New York (2014), 216
Persian Gulf, 214
Peru, 96–114, 133–50, 162, 209, 210, 216, 217, 218
  Amazonia, 14, 133–50, 162
  Corrientes River, 137, 140, 141, 146
  Fujimori administration, 99, 100
  García administration, 100
  Humala administration, 133, 134
  indigenous communities, 215
  *Junta*, 100, 103, 104
  Lima, 140, 96
  Montano Lake, 138, 141, 142, 144
  Morales administration, 98
  Pastaza River, 137, 140, 141, 147
  Tigre River, 133–50
  Velasco administration, 98
  water law, 99–100, 140
petrol. *See* oil
Philippines, 33, 34
Polanyi, Karl, 7, 10, 10–11, 88–90, 91, 107
polar bears, 12, 51, 58, 59, 62, 65, 66, 70, 118, 123–7

pollution, 8, 14, 28, 34, 35, 36, 52, 134, 135, 141, 145, 154, 162, 166, 194, 216
polluted rivers, 14, 27, 151
Polynesia, 176, 184
poverty, 15, 34, 79–80, 84, 98, 106, 108, 109
PSI (Programa Subsectorial de Irrigaciones Sierra), 107–9

quinoa, 102, 104–6, 110, 111, 112, 114

Rasmussen, Knud, 45, 46
resource extraction, 1, 151, 152, 155, 157, 160, 165, 175
  coal, 154
  mineral, 68, 125
  oil, 2, 6, 68, 133–7, 143, 146
Rio Earth Summit. *See* Earth Summit in Rio de Janeiro (1992)
rivers, 3, 13, 14, 27, 134, 136, 140, 141, 145, 146, 147, 214
Rockefeller Foundation, 33
Roosevelt, Franklin Delano, 33
Ross, John, 43, 44, 45, 52
Russia, 27, 59, 85

Salinas, Cecilia G., 10, 13, 108, 159, 162–3
Samoa, 173
Sassen, Saskia, 86
sea ice, 50, 51, 57–65, 116, 123, 124, 125, 128, 191, 196
sea level, 13, 172, 174, 179, 182, 183, 185, 219, 224
sealskin, 45, 53, 64, 124
Second World War, 7, 24, 26, 29, 32, 33, 35
Sejersen, Frank, 13, 16
Singapore, 209
slow violence, 97, 109, 135
Smith, Adam, 9
Smith Sound, 42, 46
social media, 16, 181, 205, 207, 210, 211, 216, 219

Facebook, 118, 119, 120, 129, 210, 217, 219
Instagram, 210
Twitter, 210, 217, 219
Solomon Islands, 173, 175, 178
South Africa, 158
Spain, 210
Steffen, Will, 22, 24
Stoler, Laura Ann, 42, 54
Strathern, Marilyn, 117–18
Sullivan, S., 163, 164
sustainable development, 9, 89, 100, 155, 157, 161, 217, 219
Swan, Colleen, 191, 192
Sweden, 210
Switzerland, 210, 211, 216, 218
Swyngedouw, E., 89, 189, 193

Taiwan, 33
Tanzania, 154, 158, 162
Tigre River, 133–50
*Time* (magazine), 31
Tokelau, 174, 175
Tsing, Anna, 6, 7, 8
Tuvalu, 174, 175, 176
Twitter, 210, 217, 219

Uganda, 158
UNCLOS, 171, 173, 176, 181, 184
UNESCO, 51, 52, 158, 217
United Nations (UN), 159, 171, 174, 182, 183, 184, 218
   UN REDD programme, 3, 10, 151, 152, 153, 160–62, 163, 165–6
United States of America (USA), 16, 25–6, 104, 174, 183, 206, 209, 210, 214
   DDT, 31
   military, 7, 22, 25, 26, 27, 30, 31, 32, 49, 192, 206
   Mongolia, 85
   Paris Agreement, 216

   security, 25–6
   Trump administration, 215–16
   Vietnam, 206

Verran, H., 135, 136, 145, 146
vulnerability, 5, 13, 15, 70, 100, 101, 117, 176, 188, 192
   Mongolia, 76–95

walrus, 12, 44, 45, 46, 51, 53, 60, 63, 64, 65, 66, 119, 123–7
wastelanding, 7, 24, 25, 26, 35–6
water, 136, 145, 146, 147, 154, 173, 176, 179, 182, 183, 184, 214
   access to, 15, 98, 100, 108, 109
   Arctic, 59, 60, 61, 63, 64, 71, 123, 219
   contamination, 7, 25, 135, 138, 140, 141, 142, 143
   glaciers, 50, 101, 196, 198
   Mongolia, 78, 80
   Peru, 77, 99, 100, 101, 102, 103, 108, 109, 110
   privatisation of, 100
   scarcity of, 77, 102, 108
whaling, 121, 194
Wildavsky, A., 16
Wilhite, Harold, 5, 10
World Health Organization (WHO), 141
World Trade Organization (WTO), 78, 83, 84, 86, 88, 94, 95
World War I. *See* First World War
World War II. *See* Second World War
Working Group on the Anthropocene (WGA), 7
World Bank, 33, 78, 99, 108, 164
World March for Science (2017), 216
Wyoming, 214

Zalasiewicz, J., 7
Zierler, David, 32

**The Pluto Press Newsletter**

Hello friend of Pluto!

Want to stay on top of the best radical books we publish?

Then sign up to be the first to hear about our new books, as well as special events, podcasts and videos.

You'll also get 50% off your first order with us when you sign up.

Come and join us!

Go to bit.ly/PlutoNewsletter